Is the Internet Changing the Way You Think?

ALSO BY JOHN BROCKMAN

AS AUTHOR

By the Late John Brockman
37
Afterwords
The Third Culture: Beyond the Scientific Revolution
Digerati

AS EDITOR

About Bateson
Speculations
Doing Science
Ways of Knowing
Creativity
The Greatest Inventions of the Past 2,000 Years
The Next Fifty Years
The New Humanists
Curious Minds
What We Believe but Cannot Prove
My Einstein
Intelligent Thought
What Is Your Dangerous Idea?
What Are You Optimistic About?
What Have You Changed Your Mind About?
This Will Change Everything

AS COEDITOR

How Things Are (with Katinka Matson)

IS THE INTERNET CHANGING THE WAY YOU THINK?

The Net's Impact on Our Minds and Future

Edited by John Brockman

NEW YORK • LONDON • TORONTO • SYDNEY • NEW DELHI • AUCKLAND

HARPER PERENNIAL

IS THE INTERNET CHANGING THE WAY WE THINK? Copyright © 2011 by Edge Foundation, Inc. All rights reserved. Printed in the United States of America. No part of this book may be used or reproduced in any manner whatsoever without written permission except in the case of brief quotations embodied in critical articles and reviews. For information address HarperCollins Publishers, 10 East 53rd Street, New York, NY 10022.

HarperCollins books may be purchased for educational, business, or sales promotional use. For information please write: Special Markets Department, HarperCollins Publishers, 10 East 53rd Street, New York, NY 10022.

FIRST EDITION

Designed by Justin Dodd

Library of Congress Cataloging-in-Publication data is available upon request.

ISBN 978-0-06-202044-4

11 12 13 14 15 OV/BVG 10 9 8 7 6 5 4 3 2 1

To KHM

Contents

Acknowledgments

I wish to thank Peter Hubbard of HarperCollins for his encouragement. I am also indebted to my agent, Max Brockman, who saw the potential for this book, and to Sara Lippincott for her thoughtful and meticulous editing.

Preface: The *Edge* Question

The *Edge* project was inspired by a 1971 failed art experiment. This venture was titled "The World Question Center" and was devised by the late James Lee Byars, my friend and sometime collaborator. Byars believed that to arrive at a satisfactory plateau of knowledge it was pure folly to go to Widener Library at Harvard and read 6 million books. Instead, he planned to gather the hundred most brilliant minds in the world in a room, lock them in, and have them ask one another the questions they were asking themselves. The expected result (in theory) was to be a synthesis of all thought. But it didn't work out that way. Byars identified his hundred most brilliant minds and called each of them. The result: Seventy people hung up on him.

A decade later, I picked up on the idea and founded the Reality Club, which in 1997 went online, rebranded as *Edge*. The ideas presented on *Edge* are speculative; they represent the frontiers in such areas as evolutionary biology, genetics, computer science, neurophysiology, psychology, and physics. Emerging out of these contributions is a new natural philosophy, new ways of understanding physical systems, new ways of thinking that call into question many of our basic assumptions.

For each of the anniversary editions of *Edge*, I have used the interrogative myself and asked contributors for their responses to a question that comes to me, or to one of my correspondents, in the middle of the night.

It's not easy coming up with a question. As Byars used to say: "I can answer the question, but am I bright enough to ask it?" I'm looking for questions that inspire answers we can't possibly predict. My goal is to provoke people into thinking thoughts they normally might not have.

The 2010 Edge Question

This year's question is "How is the Internet changing the way you think?" (Not "How is the Internet changing the way *we* think?" *Edge* is a conversation, and "we" responses tend to come across like expert papers, public pronouncements, or talks delivered from a stage.)

The art of a good question is to find a balance between the abstract and the personal, to ask a question that has many answers—or at least a question to which you don't know the answer. A good question encourages answers that are grounded in experience but bigger than any experience alone. I wanted *Edge*'s contributors to think about the Internet, which includes but is a much bigger subject than the Web or an application on the Internet (or searching, browsing, and so forth, which are apps on the Web). Back in 1996, computer scientist and visionary Danny Hillis pointed out: "A lot of people think the Web *is* the Internet, and they're missing something. The Web is the old media incorporated into the new medium." He enlarges on that thought in the introduction.

This year, I enlisted the aid of Hans Ulrich Obrist, curator of the Serpentine Gallery in London, and the artist April Gornik, one of the early members of the Reality Club, to help broaden the *Edge* conversation—or, rather, to bring it back to where it was in the late 1980s and early 1990s, when April gave a talk at a Reality Club meeting and discussed the influence of chaos theory on her work, and Benoit Mandelbrot showed up to discuss fractal theory. Every artist in New York City wanted to be there. What then happened was very interesting. When the Reality Club went online as *Edge*, the scientists were all on e-mail—and the artists weren't. Thus did *Edge*, surprisingly, become a science site, whereas my own background (beginning in 1965, when Jonas Mekas hired me to manage the Film-Makers' Cinematheque) was in the visual and performance arts. Gornik and Obrist have brought a number of artists into our annual colloquy.

Their responses were varied and interesting: Gornik's (with Eric Fischl) "Replacing Experience with Facsimile"; Marina Abramović, "My Perception of Time"; Stefano Boeri, "internet is wind"; Terence Koh, "a completely new form of sense"; Matthew Ritchie, "What's Missing Here?"; Brian Eno, "What I Notice"; James Croak, "Art Making Going Rural"; Raqs Media Collective, "No One Is Immune to the Storms That Shake the World"; Jonas Mekas, "I Am Not Exactly a Thinking Person—I Am a Poet"; and Ai Weiwei, who wrote, "When I'm on the Net, I Start to Think."

A new invention has emerged, a code for the collective consciousness that requires a new way of thinking. The collective externalized mind is the mind we all share. The Internet is the infinite oscillation of our collective consciouness interacting with itself. It's not about computers. It's not about what it means to be human—in fact, it challenges, renders trite, our cherished assumptions on that score. It's about thinking. Here, more than 150 *Edge* contributors—scientists, artists, creative thinkers—explore what it means to think in the new age of the Internet.

John Brockman
Publisher and Editor, *Edge*

Introduction: The Dawn of Entanglement

W. Daniel Hillis

Physicist, computer scientist; chairman, Applied Minds, Inc.; author, The Pattern on the Stone

It seems that most people, even intelligent and well-informed people, are confused about the difference between the Internet and the Web. No one has evidenced this misunderstanding more clearly than Tom Wolfe in a turn-of-the millennium essay titled "Hooking Up":

> I hate to be the one who brings this news to the tribe, to the magic Digikingdom, but the simple truth is that the Web, the Internet, does one thing. It speeds up the retrieval and dissemination of information, partially eliminating such chores as going outdoors to the mailbox or the adult bookstore, or having to pick up the phone to get hold of your stock broker or some buddies to shoot the breeze with. That one thing the Internet does and only that. The rest is Digibabble.

This confusion between the network and the services that it first enabled is a natural mistake. Most early customers of electricity believed they were buying electric lighting. That first application was so compelling that it blinded them to the bigger picture of what was possible. A few dreamers speculated that electricity would change the world, but one can imagine a nineteenth-century curmudgeon attempting to dampen their enthusiasm: "Electricity is a convenient means to light a room. That one thing the electricity does and only that. The rest is Electrobabble."

The Web is a wonderful resource for speeding up the retrieval and dissemination of information, and that, despite Wolfe's trivialization, is no small change. Yet the Internet is much more than just the Web. I would like to discuss some of the less apparent ways in which it will change us. By the Internet, I mean the global network of interconnected computers that enables, among other things, the Web. I would like to focus on applications that go beyond human-to-human communication. In the long run, these are the applications of the Internet that will have the greatest impact on who we are and how we think.

Today, most people recognize that they are using the Internet only when they are interacting with a computer screen. They are less likely to appreciate that they are using the Internet while talking on the telephone, watching television, or flying on an airplane. Some air travelers may have recently gotten a glimpse of the truth, for example, upon learning that their flights were grounded due to a router failure in Salt Lake City, but for most of them this was just another inscrutable annoyance. Most people long ago gave up trying to understand how technical systems work. This is a part of how the Internet is changing the way we think.

I want to be clear that I am not complaining about technical ignorance. In an Internet-connected world, it is almost impossible to keep track of how systems actually function. Your telephone conversation may be delivered over analog lines one day and by the Internet the next. Your airplane route may be chosen by a computer, a human being, or (most likely) some combination of both. Don't bother asking, because any answer you get is likely to be wrong.

Soon no human will know the answer. More and more decisions are made by the emergent interaction of multiple communicating systems, and these component systems themselves are constantly adapting, changing the way they work. This is the real impact of the Internet: By allowing adaptive complex systems to interoperate, the Internet has changed the way we make decisions. More and more, it is not individual humans who decide but an entangled, adaptive network of humans and machines.

To understand how the Internet encourages this interweaving of complex systems, you need to appreciate how it has changed the nature of computer programming. Back in the twentieth century, programmers could exercise absolute control within a bounded world with precisely defined rules. They were able to tell their computers exactly what to do. Today, programming usually involves linking together complex systems developed by others without understanding exactly how they work. In fact, depending upon the methods of other systems is considered poor programming practice, because it is expected that they will change.

Consider, as a simple example, a program that needs to know the time of day. In the unconnected world, computers often asked the operator to type in the time when they were powered on. They then kept track of passing time by counting ticks of an internal clock. Programmers often had to write their own program to do this, but in any case they understood exactly how it worked. Once computers became connected through the Internet, it made more sense for computers to find out the time by asking one another, so something called Network Time Protocol was invented. Most programmers are aware that it exists, but few understand it in detail. Instead, they call a library routine—a routine that queries the operating system, which automatically invokes the Network Time Protocol when required.

It would take me too long to explain the workings of Network Time Protocol and how it corrects for variable network delays and takes advantage of a partially layered hierarchy of network-connected clocks to find the time. Suffice it to say that it's complicated. Besides, I would be describing version 3 of the protocol, and your operating system is probably already using version 4. Even if you're a programmer, there's no need for you to bother to understand how it works.

Now consider a program that is directing delivery trucks to restock stores. It needs to know not just the time of day but also the locations of the trucks in the fleet, the maps of the streets, the coordinates of the warehouses, the current traffic patterns, and the inventories of the stores. Fortunately, the program can keep track of all of this changing information by connecting to other computers through the Internet. The program can also offer services to other company systems, which need to track the location of the packages, pay the drivers, and schedule maintenance of the trucks. All these systems will depend on one another to provide information, without having to understand exactly how the information is computed. These communicating systems are being constantly improved and extended, evolving in time.

Now multiply this picture a millionfold, to include not just one fleet of trucks but all the airplanes, gas pipelines, hospitals, factories, oil refineries, mines, and power plants, not to mention the salespeople, advertisers, media distributors, insurance companies, regulators, financiers, and stock traders. You will begin to perceive the entangled system that makes so many of our day-to-day decisions. Although we created it, we did not exactly design it. It evolved. Our relationship to it is similar to our relationship to our biological ecosystem. We are codependent and not entirely in control.

We have embodied our rationality within our machines and delegated to them many of our choices—and thereby created a world beyond our understanding. Our current century began on a note of uncertainty, as we worried about how our machines would handle the transition to the new millennium. Now we are attending to a financial crisis caused by the banking system having miscomputed risks, and to a debate on global warming in which experts argue not so much about the data as about what the computers predict from the data. We have linked our destinies not only to one another across the globe but also to our technology. If the theme of the Enlightenment was independence, ours is interdependence. We are now all connected, humans and machines. Welcome to the dawn of the Entanglement.

Is the Internet Changing the Way You Think?

The Bookless Library

Nicholas Carr

Author, The Shallows: What the Internet Is Doing to Our Brains

As the school year began last September, Cushing Academy, an elite Massachusetts prep school that has been around since Civil War days, announced that it was emptying its library of books. In place of the thousands of volumes that had once crowded the building's shelves, the school was installing, it said, "state-of-the-art computers with high-definition screens for research and reading," as well as "monitors that provide students with real-time interactive data and news feeds from around the world." Cushing's bookless library would become, boasted headmaster James Tracy, "a model for the twenty-first-century school."

The story gained little traction in the press—it came and went as quickly as a tweet—but to me it felt like a cultural milestone. A library without books would have seemed unthinkable just twenty years ago; today the news seems almost overdue. I've made scores of visits to libraries over the last couple of years. Every time, I've seen a greater number of people peering into computer screens than thumbing through pages. The primary role played by libraries today seems to have already shifted from providing access to printed works to providing access to the Internet. There is every reason to believe that the trend will only accelerate.

"When I look at books, I see an outdated technology," Tracy told a reporter from the *Boston Globe*. His charges would seem to agree. A sixteen-year-old student at the school took the disappearance of the library books in stride. "When you hear the word 'library,' you think of books," she said. "But very few students actually read them."

What makes it easy for an educational institution like Cushing to jettison its books is the assumption that the words in books are the same whether they're printed on paper or formed of pixels on a screen. A word is a word is a word. "If I look outside my window and I see my student reading Chaucer under a tree," said Tracy, giving voice to this common view, "it is utterly immaterial to me whether they're doing so by way of a Kindle or by way of a paperback." The medium, in other words, doesn't matter.

But Tracy is wrong. The medium does matter. It matters greatly. The experience of reading words on a networked computer, whether it's a PC, an iPhone, or a Kindle, is very different from the experience of reading those same words in a book. As a technology, a book focuses our attention, isolates us from the myriad distractions that fill our everyday lives. A networked computer does precisely the opposite. It is designed to scatter our attention. It doesn't shield us from environmental distractions; it adds to them. The words on a computer screen exist in a welter of contending stimuli.

The human brain, science tells us, adapts readily to its environment. The adaptation occurs at a deep biological level, in the way our nerve cells or neurons connect. The technologies we think with, including the media we use to gather, store, and share information, are critical elements of our intellectual environment, and they play important roles in shaping our modes of thought. That fact not only has been proved in the laboratory but also is evident from even a cursory glance at the course of intellectual history. It may be immaterial to Tracy whether a student reads from a book or a screen, but it is not immaterial to that student's mind.

My own reading and thinking habits have shifted dramatically since I first logged on to the Web fifteen years ago or so. I now do the bulk of my reading and researching online. And my brain has changed as a result. Even as I've become more adept at navigating the rapids of the Net, I have experienced a steady decay in my ability to sustain my attention. As I explained in the *Atlantic* in 2008, "What the Net seems to be doing is chipping away my capacity for concentration and contemplation. My mind now expects to take in information the way the Net distributes it: in a swiftly moving stream of particles."* Knowing that the depth of our thought is tied directly to the intensity of our attentiveness, it's hard not to conclude that as we adapt to the intellectual environment of the Net our thinking becomes shallower.

There are as many human brains as there are human beings. I expect, therefore, that reactions to the Net's influence, and hence to this year's *Edge* question, will span many points of view. Some people will find in the busy interactivity of the networked screen an intellectual environment ideally suited to their mental proclivities. Others will see a catastrophic erosion in the ability of human beings to engage in calmer, more meditative modes of thought. A great many likely will be somewhere between the extremes, thankful for the Net's riches but worried about its long-term effects on the depth of individual intellect and collective culture.

My own experience leads me to believe that what we stand to lose will be at least as great as what we stand to gain. I feel sorry for the kids at Cushing Academy.

---

* "Is Google Making Us Stupid?" *Atlantic*, July–August 2008.

The Invisible College

Clay Shirky

Social and technology network topology researcher; adjunct professor, New York University Graduate School of Interactive Telecommunications Program (ITP); author, Cognitive Surplus

The Internet has been in use by a majority of citizens in the developed world for less than a decade, but we can already see some characteristic advantages (dramatically improved access to information, very large-scale collaborations) and disadvantages (interruption-driven thought, endless distractions). It's tempting to try to judge the relative value of the network on the way we think by deciding whether access to Wikipedia outweighs access to tentacle porn or the other way around.

It is our misfortune to live through the largest increase in expressive capability in the history of the human race—a misfortune because surplus is always more dangerous than scarcity. Scarcity means that valuable things become more valuable, a conceptually easy change to integrate. Surplus means that previously valuable things stop being valuable, which freaks people out.

To make a historical analogy with the last major spread of new publishing technology, you could earn a living in 1500 simply by knowing how to read and write. The spread of those abilities in the subsequent century had the curious property of making literacy both more essential and less professional; literacy became critical at the same time as the scribes lost their jobs.

The same thing is happening with publishing. In the twentieth century, the mere fact of owning the apparatus to make something public—whether a printing press or a TV tower—made you a person of considerable importance. Today, though, publishing, in the sense of making things public, is becoming similarly deprofessionalized. YouTube is now in the position of having to *stop* eight-year-olds from becoming global publishers of video. The mere fact of being able to publish to a global audience is the new literacy—formerly valuable, now so widely available that you can't make any money with the basic capability anymore.

This shock of inclusion, where professional media give way to participation by 2 billion amateurs (a threshold we will cross this year), means that the average quality of public thought has collapsed; when anyone can say anything anytime, how could it not? If the only consequence of this influx of amateurs is the destruction of existing models for producing high-quality material, we would be at the beginning of another Dark Ages.

So it falls to us to make sure that that isn't the only consequence.

To the question "How is the Internet changing the way you think?" the right answer is "Too soon to tell." This isn't because we can't yet see some of the obvious effects but because the deep changes will be manifested only when new cultural norms shape what the technology makes possible.

To return to the press analogy, printing was a necessary but not sufficient input to the scientific revolution. The Invisible College, the group of natural philosophers who drove the original revolution in chemistry in the mid-1600s, were strongly critical of the alchemists, their intellectual forebears, who for centuries had made only fitful progress. By contrast, the Invisible College put chemistry on a sound scientific footing in a matter of a couple of decades, one of the most important intellectual transitions in the history of science. In the 1600s, though, a chemist and an alchemist used the same tools and had access to the same background. What did the Invisible College have that the alchemists didn't?

They had a culture of sharing. The problem with the alchemists wasn't that they failed to turn lead into gold; the problem was that they failed uninformatively. Alchemists were obscurantists, recording their work by hand and rarely showing it to anyone but disciples. In contrast, members of the Invisible College shared their work, describing and disputing their methods and conclusions so that they all might benefit from both successes and failures and build on one another's work.

The chemists were, to use the avant-garde playwright Richard Foreman's phrase, "pancake people." They abandoned the spiritual depths of alchemy for a continual and continually incomplete grappling with what was real, a task so daunting that no one person could take it on alone. Though the history of science we learn as schoolchildren is often marked by the trope of the lone genius, science has always been a networked operation. In this, we can see a precursor to what's possible for us today. The Invisible College didn't just use the printing press as raw capability but created a culture that used the press to support the transparency and argumentation that science relies on. We have the same opportunity.

As we know from arXiv.org, the twentieth-century model of publishing is inadequate to the kind of sharing possible today. As we know from Wikipedia, post hoc peer review can support astonishing creations of shared value. As we know from the search for Mersenne primes, whole branches of mathematical exploration are now best taken on by groups. As we know from open-source efforts such as Linux, collaboration between loosely joined parties can work at scales and over time frames previously unimagined. As we know from NASA clickworkers, groups of amateurs can sometimes replace single experts. As we know from www.patientslikeme.com, patient involvement accelerates medical research. And so on.

The beneficiaries of the system in which making things public was a privileged activity—academics, politicians, reporters, doctors—will complain about the way the new abundance of public thought upends the old order, but those complaints are like keening at a wake:

The change they are protesting is already in the past. The real action is elsewhere.

The Internet's primary effect on how we think will reveal itself only when it affects the cultural milieu of thought, not just the behavior of individual users. The members of the Invisible College did not live to see the full flowering of the scientific method, and we will not live to see what use humanity makes of a medium for sharing that is cheap, instant, and global (both in the sense of "comes from everyone" and in the sense of "goes everywhere"). We are, however, the people who are setting the earliest patterns for this medium. Our fate won't matter much, but the norms we set will.

Given what we have today, the Internet might be seen as the Invisible High School, with a modicum of educational material in an ocean of narcissism and social obsessions. We could, however, also use it as an Invisible College, the communicative backbone of real intellectual and civic change. To do this will require more than technology. It will require us to adopt norms of open sharing and participation, fitted to a world in which publishing has become the new literacy.

Net Gain

Richard Dawkins

Evolutionary biologist; emeritus Professor of the Public Understanding of Science, Oxford; author, The Greatest Show on Earth

If, forty years ago, the *Edge* question had been "What do you anticipate will most radically change the way you think during the next forty years?" my mind would have flown instantly to a then-recent article in *Scientific American* (September 1966) about Project MAC. Nothing to do with the Apple Mac, which it long predated, Project MAC was an MIT-based cooperative enterprise in pioneering computer science. It included the circle of AI innovators surrounding Marvin Minsky, but, oddly, that was not the part that captured my imagination. What really excited me, as a user of the large mainframe computers that were all you could get in those days, was something that nowadays would seem utterly commonplace: the then-astonishing fact that up to thirty people, from all around the MIT campus and even from their homes, could simultaneously log on to the same computer, simultaneously communicate with it and with each other. Mirabile dictu, the coauthors of a paper could work on it simultaneously, drawing upon a shared database in the computer, even though they might be miles apart. In principle, they could be on opposite sides of the globe.

Today that sounds absurdly modest. It's hard to recapture how futuristic it was at the time. The post-Berners-Lee world of 2010, if we could have imagined it forty years ago, would have seemed shattering. Anybody with a cheap laptop computer and a Wi-Fi connection can enjoy the illusion of bouncing dizzily around the world in full color, from a beach webcam in Portugal to a chess match in Vladivostok, and Google Earth actually lets you fly the full length of the intervening landscape, as if on a magic carpet. You can drop in for a chat at a virtual pub in a virtual town whose geographical location is so irrelevant as to be literally nonexistent (and the content of whose LOL-punctuated conversation, alas, is likely to be of a driveling fatuity that insults the technology that mediates it).

"Pearls before swine" overestimates the average chat room conversation, but it is the pearls of hardware and software that inspire me: the Internet itself and the World Wide Web, succinctly defined by Wikipedia as "a system of interlinked hypertext documents accessed via the Internet." The Web is a work of genius, one of the highest achievements of the human species, whose most remarkable quality is that it was constructed not by one individual genius such as Tim Berners-Lee or Steve Wozniak or Alan Kay, nor by a top-down company such as Sony or IBM, but by an anarchistic confederation of largely anonymous units located (irrelevantly) all over the world. It is Project MAC writ large. Suprahumanly large. Moreover, there is not one massive central computer with lots of satellites, as in Project MAC, but a distributed network of computers of different sizes, speeds, and manufacturers—a network that nobody, literally nobody, ever designed or put together but which grew, haphazardly, organically, in a way that is not just biological but specifically *ecological*.

Of course there are negative aspects, but they are easily forgiven. I've already referred to the lamentable content of many chat room conversations. The tendency to flaming rudeness is fostered by the convention—whose sociological provenance we might discuss one day—of anonymity. Insults and obscenities to which you would not dream of signing your real name flow gleefully from the keyboard when you are masquerading online as "TinkyWinky" or "FlubPoodle" or "ArchWeasel."

And then there is the perennial problem of sorting out true information from false. Fast search engines tempt us to see the entire

Web as a gigantic encyclopedia, while forgetting that traditional encyclopedias were rigorously edited and their entries composed by chosen experts. Having said that, I am repeatedly astounded by how good Wikipedia can be. I calibrate Wikipedia by looking up the few things I really do know about (and may indeed have written the entry for in traditional encyclopedias)—say, evolution or natural selection. I am so impressed by these calibratory forays that I go with some confidence to entries where I lack firsthand knowledge (which was why I felt able to quote Wikipedia's definition of the Web, above). No doubt mistakes creep in or are even maliciously inserted, but the half-life of a mistake, before the natural correction mechanism kills it, is encouragingly short. Nevertheless, the fact that the wiki concept works—even if only in some areas, such as science—flies so flagrantly in the face of all my prior pessimism that I am tempted to see it as a metaphor for all that deserves optimism about the World Wide Web.

Optimistic we may be, but there is a lot of rubbish on the Web—more than in printed books, perhaps because they cost more to produce (and, alas, there's plenty of rubbish there, too). But the speed and ubiquity of the Internet actually help us to be on our critical guard. If a report on one site sounds implausible (or too plausible to be true), you can quickly check it on several more. Urban legends and other viral memes are helpfully cataloged on various sites. When we receive one of those panicky warnings (often attributed to Microsoft or Symantec) about a dangerous computer virus, we do *not* spam it to our entire address book but instead Google a key phrase from the warning itself. It usually turns out to be, say, Hoax Number 76, its history and geography having been meticulously tracked.

Perhaps the main downside of the Internet is that surfing can be addictive and a prodigious time waster, encouraging a habit of butterflying from topic to topic rather than attending to one thing at a time. But I want to leave negativity and naysaying and end with some speculative—perhaps more positive—observations. The unplanned worldwide unification that the Web is achieving (a science fiction enthusiast might discern the embryonic stirrings of a new life-form) mirrors the evolution of the nervous system in multicellular animals. A certain school of psychologists might see it as mirroring the development of each individual's personality, as a fusion among split and distributed beginnings in infancy.

I am reminded of an insight that comes from Fred Hoyle's science fiction novel *The Black Cloud*. The cloud is a superhuman interstellar traveler whose "nervous system" consists of units that communicate with one another by radio—orders of magnitude faster than our puttering nerve impulses. But in what sense is the cloud to be seen as a single individual rather than a society? The answer is that interconnectedness that is sufficiently fast blurs the distinction. A human society would effectively become one individual if we could read one another's thoughts through direct, high-speed, brain-to-brain transmission. Something like that may eventually meld the various units that constitute the Internet.

This futuristic speculation recalls the beginning of my essay. What if we look forty years into the future? Moore's Law will probably continue for at least part of that time, enough to wreak some astonishing magic (as it would seem to our puny imaginations if we could be granted a sneak preview today). Retrieval from the communal exosomatic memory will become dramatically faster, and we shall rely less on the memory in our skulls. At present, we still need biological brains to provide the cross-referencing and association, but more sophisticated software and faster hardware will increasingly usurp even that function.

The high-resolution color rendering of virtual reality will improve to the point where the distinction from the real world becomes unnervingly hard to notice. Large-scale communal games such as Second Life will become disconcertingly addictive to many ordinary people who understand little of what goes on in the engine room.

And let's not be snobbish about that. For many people around the world, "first life" reality has few charms, and, even for those more fortunate, active participation in a virtual world is more intellectually stimulating than the life of a couch potato slumped in idle thrall to *Big Brother.* To intellectuals, Second Life and its souped-up successors will become laboratories of sociology, experimental psychology, and their successor disciplines yet to be invented and named. Whole economies, ecologies, and perhaps personalities will exist nowhere other than in virtual space.

Finally, there may be political implications. Apartheid South Africa tried to suppress opposition by banning television and eventually had to give up. It will be more difficult to ban the Internet. Theocratic or otherwise malign regimes, such as Iran and Saudi Arabia today, may find it increasingly hard to bamboozle their citizens with their evil nonsense. Whether, on balance, the Internet benefits the oppressed more than the oppressor is controversial and at present may vary from region to region (see, for example, the exchange between Evgeny Morozov and Clay Shirky in *Prospect*, November–December 2009).

It is said that Twitter played an important part in the unrest surrounding the election in Iran in 2009, and news from that faith pit encouraged the view that the trend will be toward a net positive effect of the Internet on political liberty. We can at least hope that the faster, more ubiquitous, and above all cheaper Internet of the future may hasten the long-awaited downfall of ayatollahs, mullahs, popes, televangelists, and all who wield power through the control (whether cynical or sincere) of gullible minds. Perhaps Tim Berners-Lee will one day earn the Nobel Peace Prize.

Let Us Calculate

Frank Wilczek

Physicist, MIT; 2004 Nobel laureate in physics; author, The Lightness of Being: Mass, Ether, and the Unification of Forces

Apology: The question "How is the Internet changing the way you think?" is a difficult one for me to answer in an interesting way. The truth is, I use the Internet as an appliance, and it hasn't profoundly changed the way I think—at least not yet. So I've taken the liberty of interpreting the question more broadly, as "How should *the Internet, or its descendants, affect how people like me think?"*

> If controversies were to arise, there would be no more need of disputation between two philosophers than between two accountants. For it would suffice to take their pencils in their hands, to sit down to the slates, and to say to each other (with a friend as witness, if they liked): "Let us calculate."
>
> —Leibniz (1685)

Clearly Leibniz was wrong here, for without disputation philosophers would cease to be philosophers. And it is difficult to see how any amount of calculation could settle, for example, the question of free will. But if we replace, in Leibniz's visionary program, "sculptors of material reality" for "philosophers," then we arrive at an accurate description of an awesome opportunity—and an unanswered challenge—that faces us today. This opportunity began to take shape roughly eighty years ago, as the equations of quantum theory reached maturity.

> The underlying physical laws necessary for the mathematical theory of a large part of physics and the whole of chemistry are thus completely known, and the difficulty is only that the exact

application of these laws leads to equations much too complicated to be soluble.
— P. A. M. Dirac (1929)

A lot has happened in physics since Dirac's 1929 declaration. Physicists have found new equations that reach into the heart of atomic nuclei. High-energy accelerators have exposed new worlds of unexpected phenomena and tantalizing hints of nature's ultimate beauty and symmetry. Thanks to that new fundamental understanding, we understand how stars work and how a profoundly simple but profoundly alien fireball evolved into the universe we inhabit today. Yet Dirac's bold claim holds up: While the new developments provide reliable equations for objects smaller and conditions more extreme than we could handle before, they haven't changed the rules of the game for ordinary matter under ordinary conditions. On the contrary, the triumphant march of quantum theory far beyond its original borders strengthens our faith in its soundness.

What even Dirac probably did not foresee, and what transforms his philosophical reflection of 1929 into a call to arms today, is that the limitation of being "much too complicated to be soluble" could be challenged. With today's chips and architectures, we can start to solve the equations for chemistry and materials science. By orchestrating the power of billions of tomorrow's chips, linked through the Internet or its successors, we should be able to construct virtual laboratories of unprecedented flexibility and power. Instead of mining for rare ingredients, refining, cooking, and trying various combinations scattershot, we will explore for useful materials more easily and systematically, by feeding multitudes of possibilities, each defined by a few lines of code, into a world-spanning grid of linked computers.

What might such a world grid discover? Some not unrealistic possibilities: friendlier high-temperature superconductors that would enable lossless power transmission, levitated supertrains, and computers that aren't limited by the heat they generate; superefficient photovoltaics and batteries that would enable cheap capture and flexible use of solar energy and wean us off carbon burning; superstrong materials that could support elevators running directly from Earth to space.

The prospects we can presently foresee, exciting as they are, could be overmatched by discoveries not yet imagined. Beyond technological targets, we can aspire to a comprehensive survey of physical reality's potential. In 1964, Richard Feynman posed this challenge: "Today, we cannot see whether Schrödinger's equation contains frogs, musical composers, or morality—or whether it does not. We cannot say whether something beyond it like God is needed, or not. And so we can all hold strong opinions either way."

How far can we see today? Not all the way to frogs or to musical composers (at least not good ones), for sure. In fact, only very recently did physicists succeed in solving the equations of quantum chromodynamics (QCD) to calculate a convincing proton, by using the fastest chips, big networks, and tricky algorithms. That might sound like a paltry beginning, but it's actually an encouraging show of strength, because the equations of QCD are much more complicated than the equations of quantum chemistry. And we've already been able to solve those more tractable equations well enough to guide several revolutions in the material foundations of microelectronics, laser technology, and magnetic imaging. But all these computational adventures, while impressive, are clearly warm-up exercises. To make a definitive leap into artificial reality, we'll need both more ingenuity and more computational power.

Fortunately, both could be at hand. The SETI@home project has enabled people around the world to donate their idle computer time to sift radio waves from space, advancing the search for extraterrestrial intelligence. In connection with the Large Hadron Collider (LHC) project, CERN—where, earlier, the World Wide

Web was born—is pioneering the GRID computer project, a sort of Internet on steroids that will allow many thousands of remote computers and their users to share data and allocate tasks dynamically, functioning in essence as one giant brain. Only thus can we cope—barely!—with the gush of information that collisions at the LHC will generate. Projects like these are the shape of things to come.

Pioneering programs allowing computers to play chess by pure calculation debuted in 1958; they rapidly become more capable, beating masters (1978), grandmasters (1988), and world champions (1997). In the later steps, a transition to massively parallel computers played a crucial role. Those special-purpose creations are mini-Internets (actually mini-GRIDs), networking dozens or a few hundred ordinary computers. It would be an instructive project today to set up a SETI@home-style network or a GRID client that could beat the best stand-alones. Players of this kind, once created, would scale up smoothly to overwhelming strength, simply by tapping into ever larger resources.

In the more difficult game of calculating quantum reality, we, with the help of our silicon friends, currently play like weak masters. We know the rules and make some good moves, but we often substitute guesswork for calculation, we miss inspired possibilities, and we take too long doing it. To improve, we'll need to make the dream of a world GRID into a working reality. To prune the solution space, we'll need to find better ways of parceling out subtasks in ways that don't require intense communication, better ways of exploiting the locality of the underlying equations, and better ways of building in physical insight. These issues have not received the attention they deserve, in my opinion. Many people with the requisite training and talent feel it's worthier to discover new equations, however esoteric, than to solve equations we already have, however important their application.

People respond to the rush of competition and the joy of the hunt. Some well-designed prizes for milestone achievements in the simulation of matter could have a substantial effect by focusing attention and a bit of glamour toward this tough but potentially glorious endeavor. How about, for example, a prize for calculating virtual water that boils at the same temperature as real water?

The Waking Dream

Kevin Kelly

Editor-at-large, Wired; *author,* What Technology Wants

We already know that our use of technology changes how our brains work. Reading and writing are cognitive tools that change the way in which the brain processes information. When psychologists use neuroimaging technology such as MRI to compare the brains of literates and illiterates working on a task, they find many differences—and not just when the subjects are reading.

Researcher Alexandre Castro-Caldas discovered that the brain's interhemispheric processing was different for those who could read and those who could not. A key part of the corpus callosum was thicker in literates, and "the occipital lobe processed information more slowly in [individuals who] learned to read as adults compared to those [who] learned at the usual age."* Psychologists Feggy Ostrosky-Solís, Miguel Arellano García, and Martha Peréz subjected literates and illiterates to a battery of cognitidve tests while measuring their brain waves and concluded that "the acquisition of reading and writing skills has changed the brain organization of cognitive activity in general . . . not only in language but also in visual perception, logical reasoning, remembering strategies, and formal operational thinking."†

If alphabetic literacy can change how we think, imagine how Internet literacy and ten hours a day in front of one kind of screen or another is changing our brains. The first generation to grow up screen

---

* "Targeting Regions of Interest for the Study of the Illiterate Brain," *International Journal of Psychology* 39, 1 (2004): 5–17.

† "Can Learning to Read and Write Change the Brain Organization? An Electrophysiological Study," *International Journal of Psychology* 39, 1 (2004): 27–35.

literate is just reaching adulthood, so we have no scientific studies of the full consequence of ubiquitous connectivity, but I have a few hunches based on my own behavior.

When I do long division, or even multiplication, I don't try to remember the intermediate numbers; long ago I learned to write them down. Because of paper and pencil, I am "smarter" in arithmetic. Similarly, I now no longer to try remember facts, or even where I found the facts. I have learned to summon them on the Internet. Because the Internet is my new pencil and paper, I am "smarter" in factuality.

But my knowledge is now more fragile. For every accepted piece of knowledge I find, there is, within easy reach, someone who challenges the fact. Every fact has its antifact. The Internet's extreme hyperlinking highlights those antifacts as brightly as the facts. Some antifacts are silly, some are borderline, and some are valid. You can't rely on experts to sort them out, because for every expert there is an equal and countervailing antiexpert. Thus anything I learn is subject to erosion by these ubiquitous antifactors.

My certainty about *anything* has decreased. Rather than importing authority, I am reduced to creating my own certainty—not just about things I care about but about anything I touch, including areas about which I can't possibly have any direct knowledge. That means that, in general, I assume more and more that what I know is wrong. We might consider this state perfect for science, but it also means I'm more likely to have my mind changed for incorrect reasons. Nonetheless, the embrace of uncertainty is one way my thinking has changed.

Uncertainty is a kind of liquidity. I think my thinking has become more liquid. It is less fixed, like text in a book, and more fluid, like, say, text in Wikipedia. My opinions shift more. My interests rise and fall more quickly. I am less interested in Truth with a capital *T* and more interested in truths, plural. I accord the subjective an important role in assembling the objective from many data points. The incremental, plodding progress of imperfect science seems the only way to know anything.

While hooked into the network of networks, I feel as though I'm a network myself, trying to achieve reliability from unreliable parts. And in my quest to assemble truths from half-truths, nontruths, and some other truths scattered in the flux (this creation of the known is now *our* job and not the job of authorities), I find my mind attracted to fluid ways of thinking (scenarios, provisional beliefs) and fluid media such as mashups, Twitter, and search. But as I flow through this slippery Web of ideas, it often feels like a waking dream.

We don't really know what dreams are for—only that they satisfy some fundamental need. Someone watching me surf the Web, as I jump from one suggested link to another, would see a daydream. Today I was in a crowd of people who watched a barefoot man eat dirt, then the face of a boy who was singing began to melt, then Santa burned a Christmas tree, then I was floating inside a mud house on the very tippy-top of the world, then Celtic knots untied themselves, then a guy told me the formula for making clear glass, then I was watching myself back in high school riding a bicycle. And that was just the first few minutes of my day on the Web this morning. The trancelike state we fall into while following the undirected path of links may be a terrible waste of time—or, like dreams, it might be a productive waste of time. Perhaps we are tapping into our collective unconscious in a way that we cannot when we are watching the directed stream of TV, radio, and newspapers. Maybe click-dreaming is a way for all of us to have the same dream, independent of what we click on.

This waking dream we call the Internet also blurs the difference between my serious thoughts and my playful thoughts—or, to put it more simply, I can no longer tell online when I'm working and when I'm playing. For some people, the disintegration between these two realms marks all that is wrong with the Internet: it is the high-priced waster of time, it breeds trifles. On the contrary, I cherish a good wasting of time as a necessary precondition for creativity. More important, I believe that the conflation of play and work, of thinking hard and thinking playfully, is one the greatest things the Internet has done.

In fact, the propensity of the Internet to diminish our attention is overrated. I do find that smaller and smaller bits of information can command the full attention of my overeducated mind. And it is not just me; everyone reports succumbing to the lure of fast, tiny interruptions of information. In response to this incessant barrage of bits, the culture of the Internet has been busy unbundling larger works into minor snippets for sale. Music albums are chopped up and sold as songs; movies become trailers, or even smaller video snips. (I find that many trailers are better than their movie.) Newspapers become Twitter posts. Scientific papers are served up in snippets on Google. I happily swim in this rising ocean of fragments.

While I rush into the Net to hunt for these tidbits, or to surf on its lucid dream, I've noticed a different approach to my thinking. My thinking is more active, less contemplative. Rather than beginning to investigate a question or hunch by ruminating aimlessly, my mind nourished only by my ignorance, I start doing things. I immediately, instantly *go*.

I go looking, searching, asking, questioning, reacting to data, leaping in, constructing notes, bookmarks, a trail, a start of making something mine. I don't wait. Don't have to wait. I act on ideas first now, instead of thinking on them. For some folks, this is the worst of the Net—the loss of contemplation. Others feel that all this frothy activity is simply stupid busywork, spinning of wheels, illusory action.

I ask myself, "Compared to what?" Compared to the passive consumption of TV or sucking up bully newspapers, or to merely sitting at home going in circles, musing about stuff in my head without any new inputs? I find myself much more productive by acting first.

The emergence of blogs and Wikipedia are expressions of this same impulse, to act (write) first and think (filter) later. To my eye, the hundreds of millions people online this very minute are not wasting time with silly associative links but are engaged in a more productive way of thinking than the equivalent hundreds of millions people were fifty years ago.

This approach does encourage tiny bits—but surprisingly, at the same time, it allows us to give more attention to works that are far more complex, bigger, and more complicated than ever before. These new creations contain more data and require more attention over longer periods, and they are more successful as the Internet expands. This parallel trend is less visible at first, because of a common shortsightedness that equates the Internet with text.

To a first approximation, the Internet is words on a screen—Google, papers, blogs. But this first glance ignores the vastly larger underbelly of the Internet—moving images on a screen. People (and not just young kids; I include myself) no longer go to books and text first. If people have a question, they head first for YouTube. For fun, we go to online massive games, or catch streaming movies, including factual videos (documentaries are in a renaissance). New visual media are stampeding onto the Net. This is where the Internet's center of attention lies, not in text alone. Because of online fans, streaming on demand, rewinding at will, and all the other liquid abilities of the Internet, directors started creating movies—such as *Lost* and *The Wire*—that were more than a hundred hours long.

These epics had multiple, interweaving plot lines, multiple protagonists, and an incredible depth of characters, and they demanded sustained attention that not only was beyond that required by previous TV and ninety-minute movies but also would have shocked Dickens and other novelists of yore. ("You mean they could follow all that, and then want more? Over how many years?") I would never have believed myself capable of enjoying such complicated stories or caring about them enough to put in the time. My attention has grown. In a similar way, the depth, complexity, and demands of games can equal those marathon movies or any great book.

But the most important way the Internet has changed the direction of my attention, and thus my thinking, is that it has become one thing. It may look as though I'm spending endless nanoseconds on a series of tweets, endless microseconds surfing Web pages or wandering among channels, endless minutes hovering over one book snippet after another—but in reality I'm spending ten hours a day paying attention to the Internet. I return to it after a few minutes, day after day, with essentially full-time attention. As do you.

We are developing an intense, sustained conversation with this large thing. The fact that it's made up of a million loosely connected pieces is distracting us. The producers of Websites, the hordes of commenters online, and the movie moguls reluctantly letting us stream their movies don't believe that their products are mere pixels in a big global show, but they are. It is one thing now: an intermedia with 2 billion screens peering into it. The whole ball of connections—including all its books, pages, tweets, movies, games, posts, streams—is like one vast global book (or movie, etc.), and we are only beginning to learn how to read it. Knowing that this large thing is there, and that I am in constant communication with it, has changed how I think.

To Dream the Waking Dream in New Ways

Richard Saul Wurman

Architect, cartographer; founder, TED Conference; author, 33: Understanding Change and the Change in Understanding

In the beginning, I drew a circle in the sand with a stick.

The pencil changed how I thought. The ballpoint pen changed how I thought. My increasing vocabulary changed what I could think of.

The telephone allowed me the fiction of being in remote places.

My first Sharp Wizard extended my memory.

Television continues to increase my understanding at an explosive rate.

Each and every modality allows for the ability of your mind to dream the waking dream in new ways.

Louis Kahn designed his buildings using vine charcoal on yellow "trash" paper. This allowed him to draw over and over the same drawing and smudge it out with the ball of his hand. This affected his designs.

Frank Gehry dreams in scrawls and crushed paper, and they transform magically into reality.

Each modality changes even what you can even think of. The Internet is just one big step along the way to flying through understanding and the invention of patterns.

It's a good one.

Tweet Me Nice

Ian Gold and Joel Gold

Ian Gold: *Neuroscientist; Canada Research Chair in Philosophy and Psychiatry, McGill University*

Joel Gold: *Psychiatrist; clinical assistant professor of psychiatry, New York University School of Medicine*

The social changes the Internet is bringing about have changed the way the two of us think about madness. The change in our thinking started, strangely enough, with reflections on Internet friends. The number of your Facebook friends, like the make of the car you drive, confers a certain status. It is not uncommon for someone to have virtual friends in the hundreds, which seems to show, among other things, that the Internet is doing more for our social lives than wine coolers or the pill.

In the days before Facebook and Twitter, time placed severe constraints on friendship. Even the traditional Christmas letter, now a fossil in the anthropological museum, couldn't be stamped and addressed 754 times by anybody with a full-time job. Technology has transcended time and made the Christmas letter viable again, no matter how large one's social circle. Ironically, electronic social networking has made the Christmas letter otiose; your friends hardly need an account of the year's highlights when they can be fed a stream of reports on the day's, along with your reflections on logical positivism or Lady Gaga.

It's hard to doubt that more friends are a good thing, friendship being among life's greatest boons. As Aristotle put it, "Without friends no one would choose to live, though he had all other goods." But of course friends are only as good as they are genuine, and it is hard to know what to think about Facebook friends. This familiar idea was made vivid to us recently by a very depressed young woman who came to see one of us for the first time. Among the causes of her depression, she said, was that she had no friends. Sitting on her psychiatrist's couch, desperately alone, she talked; and while she talked, she tweeted. Perhaps she was simply telling her Twitter friends that she was in a psychiatrist's office; perhaps she was telling them that she was talking to her psychiatrist about having no real friends; and perhaps—despite her protestations to the contrary—she was getting some of friendship's benefits from a virtual community. In the face of this striking contrast between the real and the virtual, however, it's hard not to think that a Facebook or Twitter friend isn't quite what Aristotle had in mind.

Still, one probably shouldn't make too much of this. Many of the recipients of Christmas letters wouldn't have been counted as friends in Aristotle's sense, either. There is a distinction to be made between one's friends and one's social group, a much larger community, which might include the Christmas letter people, the colleagues one floor below, or the family members you catch up with at Bar Mitzvahs and funerals. Indeed, the Internet is creating a hybrid social group that includes real friends and friends of friends who are little more than strangers. Beyond these, many of us are also interacting with genuine strangers in chat rooms, virtual spaces, and Second Life.

In contrast with friendship, however, an expanded social group is unlikely to be an unalloyed good, because it is hardly news that the people in our lives are the sources of not only our greatest joys but also our most profound suffering. The sadistic boss can blight an existence, however full it may be of affection from others, and the sustaining spouse can morph into That Cheating Bastard. A larger social group is thus a double-edged sword, creating more opportunities for human misery as well as satisfaction. A hybrid social group that includes near-strangers and true strangers may also open the door to real danger.

The mixed blessings of social life seem to have been writ large in our evolutionary history. The last time social life expanded as significantly as it has in the last couple of years was before there were any humans. The transition from nonprimates to primates came with an expansion of social groups, and many scientists now think that the primate brain evolved under the pressures of this novel form of social life. With a larger social group, there are more opportunities for cooperation and mutual benefit, but there are also novel threats. Members of a social group will each get more food if they hunt together, for example, than they would by hunting alone, but they also expose themselves to free riders, who take without contributing. With a larger social group, the physical environment is more manageable, but deception and social exploitation emerge as new dangers. Since both cooperation and competition are cognitively demanding, those with bigger brains—and the concomitant brainpower—will have the advantage in both. The evolution of human intelligence thus may have been driven primarily by the kindness or malice of others.

Some of the best evidence for this idea is that there is a relation in primates between brain size (more precisely, relative neocortical volume) and the size of the social group in which the members of the species live: bigger brain, bigger group. Plotting social group as a function of brain size in primates allows us to extrapolate to humans. The evolutionary anthropologist Robin Dunbar calculated that the volume of the human cortex predicts a social group of 150—about the size of the villages that would have constituted our social environment for a great deal of evolutionary time and which can still be found in "primitive" societies.*

How might one test this hypothesis? In nonhuman primates, membership in a social group is typically designated by mutual grooming.

---

* Robin Dunbar, *How Many Friends Does One Person Need: Dunbar's Number and Other Evolutionary Quirks* (London: Faber and Faber, 2010).

Outside of hairdressing colleges and teenage girls' sleep-overs, this isn't a very useful criterion for humans. But the Christmas letter (or card) does better. Getting a Christmas card is a minimal indicator of membership in someone's social group. In an ingenious experiment, Dunbar asked subjects to keep a record of the Christmas cards they sent. Depending on how one counted, the number of card recipients was somewhere between 125 and 154, just about the right number for our brains. It appears, then, that over the course of millions of years of human history our brains have been tuned to the social opportunities and threats presented by groups of 150 or so. The Internet has turned the human village into a megalopolis and has thus inaugurated what might be the biggest sea change in human evolution since the primeval campfires.

We come at last to madness. Psychiatry has known for decades that the megalopolis—indeed, a city of any size—breeds psychosis. In particular, schizophrenia, the paradigm of a purely biological mental illness, becomes more prevalent as city size increases, even when the city is hardly more than a village. This is not because mental illness in general becomes more common in cities; nor is it true that people who are psychotic tend to drift toward cities or stay in them. In creating much larger social groups for ourselves, ranging from true friends to near strangers, could we be laying the ground for a pathogenic virtual city in which psychosis will be on the rise? Or will Facebook and Twitter draw us closer to friends in Aristotle's sense, who can act as psychic prophylaxis against the madness-making power of others? Whatever the effects of the Internet on our inner lives, it seems clear that in changing the structure of our outer lives—the lives intertwined with those of others—the Internet is likely to be a more potent shaper of our minds than we have begun to imagine.

The Dazed State

Richard Foreman

Playwright and director; founder, the Ontological-Hysteric Theater

How is the Internet changing the way I think? But what is this "thinking" that I assume I do along with everybody else? Probably there is no agreement about what thinking consists of. But I certainly do not believe that "gathering information" is thinking, and that has obviously been an activity that has expanded and sped up as a result of the Internet. For me, to think is to withdraw from gathered information into a blankness, within which something arises, pops out, is born.

Of course it will be maintained that what pops out may have its roots—may be conditioned—by many factors in my experiential past. Nevertheless, whereas the Internet swamps us in "connectedness" and "facts," it is only in the withdrawal from those that I claim a space for thinking.

So in one sense the Internet expands the arena within which thinking may resonate, and so perhaps the thinking is thereby attuned somewhat differently. But I am one of those who believe that while it is clearly life-changing, it is no way—if you will—soul-changing. Accessing the ever expanding, ever faster Internet means a life that is changing as it becomes the life of a surfer (just as life might change if you moved to a California beach community). You become more and more agile, balancing on top of the flow, leaping from hyperlink to hyperlink, giving your mental environment a certain shape based on those chosen jumps.

But the Internet sweeps you away from where and what you were. So instead of filling you with the fire to dig deeper into the magic bottomless source that is the self, it lets you drift into the dazed state of having everything at your fingertips—which you use in order to caress the world, of course, but only the world as it assumes the shape of the now manifest, rather than the world of the still unimaginable.

So, even though I spend *lots* of time on the Internet (fallen, pancake person that I am), I can't help being reminded of the Greek philosopher who attributed his long life to avoiding dinner parties. If only I could avoid the equally distracting Internet, which, in its promise of connectedness and expanded knowledge, is really a substitute social phenomenon.

The "entire world" the Internet seems to offer harmonizes strangely with the apple from the Tree of Knowledge offered to Eve—ah, we don't believe in those old myths? (I guess one company guru did.)

Well, the only hope I see hovering in the Neverland (now real) where the Internet does its work of feeding smart people amphetamines and "dumb" people tranquilizers is that the expanding puddle of boiling, bubbling hot milk will eventually coagulate and a new, unforeseen pattern will emerge out of all that activity that thought it was aiming at a particular goal but (as is usual with life) was really headed someplace else, which nobody knew about.

That makes it sound like the new mysticism for a new Dark Ages. Well, we've already bitten the apple. Good luck to those much younger than I am, who may be around to see either the new Heaven or the new Hell.

What's Missing Here?

Matthew Ritchie

Artist

Supposedly the Internet was invented at CERN. If CERN is really responsible for this infinitely large filing cabinet, filled to bursting by lunatics, salesmen, hobbyists, and pornographers and fitting into my pocket folded up like Masefield's Box of Delights, then CERN has posed an even larger threat to the world than its fabled potential production of black holes.

Nonetheless, I use the Internet—or does it use me? Is it a new cultural ecology, an ecology of mind? If it is, who are the real predators, and who is being eaten online? Is it me?

Once I longed to create an interface that would simulate my interaction with the real world. Now I realize that the interface I want is the real world. Can the Internet give me that back?

Is it an archive? I can learn a new idea every day on the Internet. I have learned about many old ideas and many false ideas. I have read many obvious lies. This ability to indefinitely sustain a lie is celebrated as freedom. Denial enters stage left, cloaked as skepticism. We need a navigation system we can trust. Someday soon we'll need our twentieth-century experts and interpreters to be replaced by twenty-first-century creator-pilots.

Is it an open system? It seems impossible to find out on the Internet what it really costs the planet to sustain the Internet and its toys, what it costs our culture to think, to play, to fondle and adore itself. Seven of the world's largest corporations own all the routers and cables. Everyone pays the ferryman.

Is it liberating? The old, the poor, and the uneducated are locked out. Everyone else is locked in; studies show mass users locked in reversed and concentric learning patterns, seeking only the familiar—even (perhaps especially) if novelty is their version of the same old thing. As a shared space, it is a failure, celebrating only those who obey its rules. We sniff out our digital blazes, following the circular depletion of our own curiosity reservoirs. We are running out of selves.

Is it really just about communication? To travel is to enter a world of monastic chimes and insectile clicks, as unloved cell phone chatter is replaced by mobile anchorites locked in virtual communion with their own agendas and prejudices, cursing when their connections fail and they are returned to the real, immediate world. But unplugging only returns us, and them, to a space-in-waiting, designed and ordered by the same system.

Is it a new space? If this is true, then immediately I am drawn to the implied space inevitably also being created, the anti-Net. If it's a new space, how big are we when we are online? But what's really missing here? Meaning, touch, time, and place are what's missing here. We need a holographic rethinking of scale and content.

But like you, I'm back every day, "collaborating," as they say. Because there is something being built, or building itself, in this not-yet-space. Perhaps the Internet we know is merely a harbinger and—like Ulysses returning dirty, false, and lame—it will truly reveal itself only when we're ready. Perhaps it will unfold itself soon and help us bring the real ecology back to life, unveil the conspiracies, shatter the mirrors, tear down the walls, rejoice, and bring forth the promise that is truly waiting in us, waiting only for its release. I'm ready now.

Power Corrupts

Daniel C. Dennett

Philosopher; University Professor, codirector, Center for Cognitive Studies, Tufts University; Breaking the Spell: Religion as a Natural Phenomenon

We philosophers don't agree about much, but one simple slogan that just about everybody accepts is "*ought* implies *can*." You aren't obliged to do something impossible (for you). In the past, this handily excused researchers from scouring the world's libraries for obscure works that might have anticipated their apparently novel and original discoveries, since life is short and the time and effort that would have to be expended to do a thorough job of canvassing would be beyond anybody's means. Not anymore. Everybody has all-but-free and all-but-instantaneous access to the world's archives on just about every topic. A few seconds with Google Scholar can give you a few hundred more peer-reviewed articles to check out. But this is really more scholarly can-do than I want. I don't want to spend my precious research time scrolling through miles of published work, even with a well-tuned search engine! So (like everyone else, I figure), I compromise. I regret the loss of innocence imposed on me by the Internet. "I could have done otherwise, but I didn't" is the constant background refrain of all the skimpings I permit myself, all the shortcuts I take, and thus a faint tinge of guilt hangs over them all.

I also find that I am becoming a much more reactive thinker, responding—how can I do otherwise?—to a host of well-justified requests for my assistance (it will only take a few minutes) and postponing indefinitely my larger, more cumbersome projects that require a few uninterrupted hours just to get rolling. This tiny *Edge* essay is a prime example. It would be easy to resist this compression of my attention span if there weren't so many good reasons offered for taking these interruptions seriously. To date, my attempts to fend off this unwelcome trend by raising the threshold of my imperviousness have failed to keep up with the escalation. Stronger measures are called for. But do I regret the time spent writing this piece? No, on reflection I can convince myself that it may actually bring more valuable illumination to more people than a whole philosophical monograph on mereology or modal realism (don't ask). But will I ever get back to my book writing?

As Lord Acton famously said (I know—I just did a search to make sure I remembered it correctly—he said it in a letter to Bishop Mandell Creighton in 1887): "Power tends to corrupt, and absolute power corrupts absolutely." We are all today in possession of nearly absolute power in several—but not all—dimensions of thinking, and since this hugely distorts the balance between what is hard and what is easy, it may indeed corrupt us all in ways we cannot prevent.

The Rediscovery of Fire

Chris Anderson

Curator, TED Conferences, TED Talks

Amid the apocalyptic wailing over the Internet-inflicted demise of print, one countertrend deserves a hearing. The Web has allowed the reinvention of the spoken word. Thanks to an enormous expansion of low-cost bandwidth, the cost of online video distribution has fallen almost to zero. As a result, recorded talks and lectures are spreading across the Web like wildfire. They are tapping into something primal and powerful.

Before Gutenberg, we had a different technology for communicating ideas and information. It was called talking. Human-to-human speech is powerful. It evolved over millions of years, and there's a lot more happening than just the words passing from brain to brain. There's modulation, tone, emphasis, passion. And the listener isn't just listening. She's watching. Subconsciously she notes the widening of the speaker's eyes, the movement of the hands, the swaying of the body, the responses of other listeners. It all registers and makes a difference to the way the receiving brain categorizes and prioritizes the incoming information. By increasing the motivation to understand, the speaker's lasting effect on the intellectual world of the listener may be far greater than the same words in print.

Read a Martin Luther King Jr. speech and you may nod in agreement. But then track down a video of the man in action, delivering those same words in front of an energized crowd. It's a wholly different experience. You feel the force of the words. Their intent seems clearer, more convincing. You end up motivated, inspired. And so throughout history, when people have wanted to persuade, they have gathered a crowd together and made their case, often with startling effect.

If nonverbal communication has a far greater impact than verbal, how did books catch on? Simple. They offered scale. It might be harder to explain and inspire via the printed page, but if you could, tens of thousands could benefit. And so we ended up with a mass communication culture in which, for a while, books and other printed media were the stars. And surprisingly, although radio and television could have reopened the door to spoken persuasion, they largely ignored the opportunity. In the increasingly frenetic battle for attention—and constrained by economic models that required mass audiences—victory went to entertainment, news, gossip, drama, and sports. "Talking heads" were regarded as bad television, and little effort went into figuring out how to present them in an interesting way.

Meanwhile in the academic world, the emphasis was on papers and research—and somehow teaching schedules settled on painfully long lectures as the default unit of verbal communication. Man in coat behind lectern reading notes, while audience snoozes. All the intellectual brilliance in the world matters not a whit if the receiving brains can't register it as interesting.

Our ancestors would have been appalled. They knew better. Picture a starry night outside a village, in one of the ancient cradles of civilization. The people gather. The fire is lit. The drums beat. The dancers sway. A respected elder hushes the crowd. His face illuminated by flickering flames, he begins telling a story, his voice rising as the drama builds. The meaning of the story becomes apparent. The gathered crowd roars its approval. All of those present have understood something new. And more than that, they have felt it. They will act on it.

This is a scene that has played countless times in our evolutionary history. It's not unreasonable to think that our brains are fine-tuned to respond to evocative speech delivered in a powerful theatrical setting by a talented speaker.

And now the Web is making it possible for such speakers to do what print authors have been doing for centuries: reach a mass audience. What's more, the online explosion in serious talks could rectify the Web-inflicted damage to book authors' bank balances and thereby make it possible to continue making a living as a contributor to the world's intellectual commons. For one thing, when a talk goes viral, it boosts the author's book sales and generates new connections, contracts, and consultancies. Significantly, it also creates demand for the author's paid speaking appearances. Those $20K speaker fees soon add up. (An underreported effect of the increase in our time online is a growing craving for live experience. You can see it in the music industry, where all the revenue is moving away from album sales toward live performances. It's easy to imagine musicians of the future making all their music free digitally but creating unforgettable live experiences for their fans at $100 a ticket. The same may be starting to happen for book authors.)

Beyond that, there are numerous brilliant thinkers, researchers, and inventors who would never contemplate writing a book. They, too, now have the opportunity to become one of the world's teachers. Their efforts, conveyed vividly from their own mouths, will bring knowledge, understanding, passion, and inspiration to millions.

When Marshall McLuhan said, "The medium is the message," he meant, among other things, that every new medium spawns its own unexpected units of communication. In addition to the Web page, the blog, and the tweet, we are witnessing the rise of riveting online talks, long enough to inform and explain, short enough for mass impact.

The Web has allowed us to rediscover fire.

The Rise of Social Media Is Really a Reprise

June Cohen

Director of media, TED Conference; TED Talks

In the early days of the Web, when I worked at *HotWired*, I thought mainly about the new. We were of the future, those of us in that San Francisco loft—champions of new media, new tools, new thinking. But lately I've been thinking more about the old—about those aspects of human character and cognition that remain unchanged by time and technology. Over the past two decades, I've watched as the Internet changed the way we think and changed the way we live. But it hasn't changed us fundamentally. In fact, it may be returning us to the intensely social animals we evolved to be.

Every day, hundreds of millions of people use the Internet to blog, tweet, IM, and Facebook, as if it were the most natural thing in the world. And it is. The tools are new, but the behaviors come naturally. Because the rise of social media is a reprise, a return to the natural order.

When you take the long view—when you look at the Internet on an evolutionary timeline—everything we consider "old media" is actually very new. Books and newspapers became common only in the last two hundred years, radio and film in the last hundred, TV in the last fifty. If all of human history were compressed into a single twenty-four-hour day, media as we now know them emerged in the last two minutes before midnight.

Before that, for the vast majority of human history, all media were social media. Media were what happened between people. Whether you think of the proverbial campfire, around which group rituals were performed and mythologies passed on, or of simple everyday interactions (teaching, gossiping, making music, making each other laugh), media were participatory. Media were social.

So what we're seeing today isn't new. It's neither the unprecedented flowering of human potential nor the death of intelligent discourse but, rather, the correction of a historical anomaly. There was a brief period of time in the twentieth century when "media" were understood as things professionals created for others to passively consume. Collectively, we have rejected this idea.

Humans are natural-born storytellers, and media have always formed the social glue that held our communities together. But mass media in the twentieth century were so relentlessly one-way that they left room for little else. TV's lure proved so powerful, so intoxicating, and so isolating that our older, participatory traditions—storytelling, music making, simply eating together as a family—fell away. TV created a global audience but destroyed the village in the process.

Enter the Internet. As soon as the technology became available to us, we began instinctively re-creating the kinds of content and communities we evolved to crave. Our ancestors lived in small tribes, keeping their friends close and their children closer. They quickly shared information that could have life-or-death consequences. They gathered round the fire for rituals and storytelling that bonded them as a tribe. And watch us now. The first thing most of us do with a new communications technology is to gather our tribe around us—e-mailing photos to our parents, nervously friending our kids on Facebook.

And we find every way we can to participate in media, to make it a group experience: We comment on YouTube videos, vote for contestants on reality shows, turn televised news events into live theater. Think of the millions who updated their Facebook status during Barack Obama's inauguration ceremony, as if to say: "I'm here. I'm with you. I'm part of this." Our contributions may not be remarkable—they may be the written equivalent of shouting "Yay!" But then, the goal isn't to be profound, it's simply to belong.

And we share stories. We're designed to. If something surprises, delights, or disgusts us, we feel an innate urge to pass it on. The same impulse that makes Internet videos "go viral" has been spreading ideas (and jokes, and chain letters) throughout history. This ancient process is merely accelerated online and made visible, quantifiable, and—almost—predictable.

And of course we're telling our own stories, too. We read regularly about celebrity bloggers with millions of fans, or Twitter campaigns that influence world events. But the truth is that most bloggers, vloggers, tweeters, and Facebookers are talking mainly to their friends. They compare lunches, swap songs, and share the small stories of their day. They're not trying to be novelists or the *New York Times.* They're just reclaiming their place at the center of their lives. When we were handed decentralized media tools of unprecedented power, we built a digital world strikingly similar to the tribal societies and oral cultures we evolved with.

So the Internet is changing how I think about my role as a member of the media—not just as a conveyer of information but as a convener of people. The Internet is changing the way I think by making me think, at every moment, "What do I think of this? Whom do I want to tell about it?"

And the Internet has me dreaming about our distant past, which feels a lot closer than you would think.

The Internet and the Loss of Tranquility

Noga Arikha

Historian of ideas; author, Passions and Tempers: A History of the Humours

I still remember typing essays on a much loved typewriter in my first year of university. Then the first computer, the first e-mail account, the slow yet fluid entry into a new digital world that felt strangely natural. The advent of the Internet age happened progressively; we saw it develop like a child born of many brains, a protean animal whose characteristics were at once predictable and unknown. As soon as the digital sphere had become a worldwide reality, recognizable as a new era, predictions and analyses about it grew. *Edge* itself was born as the creature was still growing new limbs. The tools for research and communication about this research developed, along with new thinking about mind-machine interaction, the future of education, the impact of the Internet on texts and writing, and the issues of filtering, relevance, learning, and memory.

And then somehow the creature became autonomous, an ordinary part of our universe. We are no longer surprised, no longer engaged in so much meta-analysis. We are dependent. Some of us are addicted to this marvelous tool, this multifaceted medium that is—as predicted even ten years ago—concentrating all of communication, knowledge, entertainment, business. I, like many of us, spend so many hours before a computer screen, typing away, even when surrounded by countless books, that it is hard to say exactly how the Internet has affected me. The Internet is becoming as ordinary as the telephone. Humans are good at adapting to the technologies we create, and the Internet is the most malleable, the most human of all technologies—just as it can also alienate us from everything we've lived as before now.

I waver between these two positions—at times gratefully dependent on this marvel, at times horrified at what this dependence signifies. Too much concentrated in one place, too much accessible from one's house, the need to move about in the real world nearly nil, the rapid establishment of social networking Websites changing our relationships, the reduction of three-dimensionality to that flat screen. Rapidity, accessibility, one click for everything: Where has slowness gone, and tranquility, solitude, quiet? The world I took for granted as a child, and which my childhood books beautifully represented, jerks with the brand-new world of artificial glare and electrically created realities—faster, louder, unrelated to nature, self-contained.

The technologies we create always have an impact on the real world, but rarely has a technology had such an impact on minds. We know what's happening to those who were born after the advent of the Internet; for those, like me, who started out with typewriters, books, slowness, reality measured by geographical distance and local clocks, the emerging world is very different indeed from the world we knew.

I am of that generation for which adapting to computers was welcome and easy but for which the pre-Internet age remains real. I can relate to those who call the radio "the wireless," and I admire people in their seventies and eighties who communicate by e-mail, because they come from further away still. Perhaps the way forward would be to emphasize the teaching of history in schools, to develop curricula on the history of technology, to remind today's children that their technology, absolutely embracing as it feels, is relative and does not represent the totality of the universe. Millions of children around the world don't need to be reminded of this—they have no access to technology at all, many not even to modern plumbing—but those who do should know how to place this tool historically and politically.

As for me, I am learning how to make room for the need to slow down and disconnect without giving up my addiction to Google, e-mail, and rapidity. I was lucky enough to come from somewhere else, from a time when information was not digitized. And that is what perhaps enables me to use the Internet with a measure of wisdom.

The Greatest Detractor to Serious Thinking Since Television

Leo Chalupa

Ophthalmologist and neurobiologist, University of California, Davis

The Internet is the greatest detractor to serious thinking since the invention of television. It can devour time in all sorts of frivolous ways, from chat rooms to video games. And what better way to interrupt one's thought processes than by an intermittent stream of incoming e-mail messages? Moreover, the Internet has made interpersonal communication much more circumscribed than in the pre-Internet era. What you write today may come back to haunt you tomorrow. The brouhaha in late 2009 following the revelations of the climate scientists' e-mails is an excellent case in point.

So while the Internet provides a means for rapidly communicating with colleagues globally, the sophisticated user will rarely reveal true thoughts and feelings in such messages. Serious thinking requires honest and open communication, and that is simply untenable on the Internet by those who value their professional reputation.

The one area in which the Internet could be considered an aid to thinking is the rapid procurement of new information. But even this is more illusory than real. Yes, the simple act of typing a few words into a search engine will virtually instantaneously produce links related to the topic at hand. But the vetting of the accuracy of information obtained in this manner is not a simple matter. What one often gets is no more than abstract summaries of lengthy articles. As a consequence, I suspect that the number of downloads of any given scientific paper has little relevance to the number of times the entire article has been read from beginning to end. My advice is that if you want to do some serious thinking, then you'd better disconnect the Internet, phone, and television set and try spending twenty-four hours in absolute solitude.

The Large Information Collider, BDTs, and Gravity Holidays on Tuesdays

Paul Kedrosky

Editor, Infectious Greed; *senior fellow, Kauffman Foundation*

Three friends have told me recently that over the latest holiday they unplugged from the Internet and had big, deep thoughts. This worries me. First, three data points means it's a trend, so maybe I should be doing it. Second, I wonder if I could disconnect from the Internet long enough to have big, deep thoughts. Third, like most people I know, I worry that even if I disconnect long enough, my info-krill-addled brain is no longer capable of big, deep thoughts (which I will henceforth call BDTs).

Could I quit? At some level, it seems a silly question, like asking how I feel about taking a breathing hiatus or if on Tuesdays I would give up gravity. The Internet no longer feels involuntary when it comes to thinking. Instead it feels more like the sort of thing that when you make a conscious effort to stop doing it bad things happen. As a kid, I once swore off gravity and jumped from a barn haymow, resulting in a sprained ankle. Similarly, a good friend of mine sometimes asks fellow golfers before a swing whether they breathe in or breathe out. The next swing is inevitably horrible, as the golfer sends a ball screaming into the underbrush.

Could I quit the Internet if it meant I would have more BDTs? Sure, I suppose I could, but I'm not convinced the BDTs would occur to me. First, the Internet is, for me, a kind of internal cognition-combustion engine, something that vastly accelerates my ability to travel vast landscapes. Without it, I'd have a harder time comparing, say, theories about complexity, cell phones, and bee colony collapse rather than writing an overdue paper or counting hotel room defaults in California versus Washington State. (In case you're curious, there are roughly twice as many defaulted hotel rooms in California as there are total hotel rooms in Seattle.)

Like most people I know, I worry noisily and loudly that the Internet has made me incapable of having BDTs. I feel sure that I used to have such things, but for some reason I no longer do. Maybe the Internet has damaged me—I've informed myself to death!—to the point that I don't know what big, deep thoughts are. Or maybe the brain chemicals formerly responsible for their emergence are now doing something else. Then again, this smacks of historical romanticism, like remembering the skies as always being blue and summers as eternal when you were eight years old.

So, as much as I kind of want to believe people who say they have big, deep thoughts when they disconnect from the Web, I don't trust them. It's as though a doctor had declared himself Amish for the day and headed to the hospital by horse and buggy with a hemorrhaging patient. Granted, you could do that and some patients might even survive, but it isn't prudent or necessary. It seems, instead, a public exercise in macho symbolism—like Iggy Pop carving something in his chest, a way of bloodily demonstrating that you're different—or even a sign of outright crankishness. Look at me! I'm thinking! No Internet!

If we know anything about knowledge, about innovation, and therefore about coming up with BDTs, it is that these things are cumulative, accretive processes of happening upon, connecting, and assembling: an infinite Erector set, not just a few pretty I-beams strewn about on a concrete floor. But if BDTs were just about connecting things, then the Internet would only be mildly interesting in changing the way I think. Libraries connect things, people connect things, and connections can even happen (yes) while sitting disconnected from the Internet under an apple tree somewhere. Here's the difference: The Internet increases the speed and frequency of these connections and collisions while dropping the cost of both to near zero.

It is that combination—cheap connections plus cheap collisions—that has done violence to the way I think. It's like having a private particle accelerator on my desktop, a way of throwing things into violent juxtaposition, with the resulting collisions reordering my thinking. The result is new particles—ideas!—some of which are BDTs and many of which are nonsense. But the democratization of connections, collisions, and therefore thinking is historically unprecedented. We are the first generation to have the information equivalent of the Large Hadron Collider for ideas. And if that doesn't change the way you think, nothing will.

The Web Helps Us See What Isn't There

Eric Drexler

Engineer, molecular technologist; author, Engines of Creation

As the Web becomes more comprehensive and searchable, it helps us see what's missing in the world. The emergence of more effective ways to detect the *absence* of a piece of knowledge is a subtle and slowly emerging contribution of the Web, yet important to the growth of human knowledge. I think we all use absence detection when we try to squeeze information out of the Web. It's worth considering both how that works and how the process could be made more reliable and user-friendly.

The contributions of absence detection to the growth of shared knowledge are relatively subtle. Absences themselves are invisible, and when they are recognized (often tentatively), they usually operate indirectly, by influencing the thinking of people who create and evaluate knowledge. Nonetheless, the potential benefits of better absence detection can be measured on the same scale as the most important questions of our time, because improved absence detection could help societies blunder toward somewhat better decisions about those questions.

Absence detection boosts the growth of shared human knowledge in at least three ways:

Development of knowledge. Generally, for shared knowledge to grow, someone must invest effort to develop a novel idea into something more substantial (resulting in a blog post, a doctoral dissertation, or whatever). A potential knowledge creator may need some degree of confidence that the expected result doesn't already exist. Better absence detection can help build that confidence—or drop it to zero and abort a costly duplication.

Validation of knowledge. For shared knowledge to grow, something that looks like knowledge must gain enough credibility to be treated as knowledge. Some knowledge is born with credibility, inherited from a credible source, yet new knowledge, supported by evidence, can be discredited by arguments backed by nothing but noise. A crucial form of evidence for a proposition is sometimes the absence of credible evidence against it.

Destruction of antiknowledge. Shared knowledge can also grow through removal of antiknowledge—for example, by discrediting false ideas that displaced or discredited true ones. Mirroring validation, a crucial form of evidence against the credibility of a proposition is sometimes the absence of credible evidence *for* it.

Identifying what is absent by observation is inherently more difficult than identifying what is present, and conclusions about absences are usually substantially less certain. The very idea runs counter to the adage, being based on the principle that absence of evidence sometimes *is* evidence of absence. This can be obvious: What makes you think there's no elephant in your room? Of course, good intellectual housekeeping demands that reasoning of this sort be used with care. Perceptible evidence must be comprehensive enough that a particular absence, in a particular place, is significant: I'm not at all sure that there's no gnat in my room, and I can't be entirely sure that there's no elephant in my neighbor's yard.

Reasonably reliable absence detection through the Web requires both good search and dense information, and this is one reason why the Web becomes effective for the task only slowly, unevenly, and almost imperceptibly. Early on, an absence in the Web shows a gap in the Web; only later does an absence begin to suggest a gap in the world itself.

I think there's a better way to detect absences, one that bypasses ad hoc search by creating a public place where knowledge comes into focus. We could benefit immensely from a medium that is as good at representing factual controversies as Wikipedia is at representing factual consensus.

What I mean by this is a social software system and community much like Wikipedia—perhaps an organic offshoot—that would operate to draw forth and present what is, roughly speaking, the best evidence on each side of a factual controversy. To function well, it would require a core community that shares many of the Wikipedia norms but would invite advocates to present a far-from-neutral point of view. In an effective system of this sort, competitive pressures would drive competent advocates to participate, and incentives and constraints inherent in the dynamics and structure of the medium would drive advocates to pit their best arguments head-to-head and point by point against the other side's best arguments. Ignoring or caricaturing opposing arguments simply wouldn't work, and unsupported arguments would become more recognizable.

Success in such an innovation would provide a single place to look for the best arguments that support a point in a debate, and with these, the best counterarguments—a single place where the absence of a good argument would be good reason to think that none exists.

The most important debates could be expected to gain traction early. The science of climate change comes to mind, but there are many others. The benefits of more effective absence detection could be immense and concrete.

Knowledge Without, Focus Within, People Everywhere

David Dalrymple

Eighteen-year-old PhD student; researcher, MIT's Mind Machine Project

Filtering, not remembering, is the most important skill for those who use the Internet. The Internet immerses us in a milieu of information—not for almost twenty years has a Web user read every available page—and there's more each minute: Twitter alone processes hundreds of tweets every second, from all around the world, all visible for anyone, anywhere, who cares to see. Of course, the majority of this information is worthless to the majority of people. Yet anything we care to know—What's the function for opening files in Perl? How far is it from Hong Kong to London? What's a power law?—is out there somewhere.

I see today's Internet as having three primary, broad consequences: (1) information is no longer stored and retrieved by people but is managed externally, by the Internet; (2) it is increasingly challenging and important for people to maintain their focus in a world where distractions are available everywhere; and (3) the Internet enables us to talk to and hear from people around the world effortlessly.

Before the Internet, most professional occupations required a large body of knowledge, accumulated over years or even decades of experience. But now anyone with good critical thinking skills and the ability to focus on the important information can retrieve it on demand from the Internet instead of from her own memory. However, those with wandering minds, who might once have been able to focus by isolating themselves with their work, now often find they must do their work with the Internet, which simultaneously furnishes a panoply of unrelated information about their friends' doings, celebrity news, limericks, and millions of other sources of distraction. How well an employee can focus might now be more important than how knowledgeable he is. Knowledge was once an internal property, and focus on the task at hand could be imposed externally; with the Internet, knowledge can be supplied externally but focus must be achieved internally.

Separable from the intertwined issues of knowledge and focus is the irrelevance of geography in the Internet age. On the transmitting end, the Internet allows many types of professionals to work in any location—from their home in Long Island, from their condo in Miami, in an airport in Chicago, or even in flight on some airlines—wherever there's an Internet connection. On the receiving end, it allows for an Internet user to access content produced anywhere in the world with equal ease. The Internet also enables groups of people to assemble based on interest, rather than on geography—collaboration can take place between people in Edinburgh, Los Angeles, and Perth nearly as easily as if they lived in neighboring cities.

In the future, we'll see the development of increasingly subconscious interfaces. Already, making an Internet search is something many people do without thinking about it, like making coffee or driving a car. Within the next fifty years, I expect the development of direct neural links, making the data that are available at our fingertips today available at our synapses in the future, and making virtual reality feel more real than our current sensory perception. Information and experience could be exchanged between our brains and the network without any conscious action. And at some point knowledge may be so external that all knowledge and experience will be universally shared, and the only notion of an "individual" will be a particular focus—a point in the vast network that concerns itself only with a specific subset of the information available.

In this future, knowledge will be fully outside the individual, focus will be fully inside, and everybody's selves will truly be spread everywhere.

A Level Playing Field

Martin Rees

President, the Royal Society; professor of cosmology and astrophysics; master, Trinity College, University of Cambridge; author, Our Final Century: The 50/50 Threat to Humanity's Survival

In 2002, three Indian mathematicians—Manindra Agrawal and two of his students, Neeraj Kayal and Nitin Saxena—invented a faster algorithm for factoring large numbers, an advance that could be crucial for code breaking. They posted their results on the Web. Such was the interest that within just a day 20,000 people had downloaded the work, which became the topic of hastily convened discussions in many centers of mathematical research around the world.

This episode—offering instant global recognition to two young Indian students—contrasts starkly with the struggles of a young Indian genius a hundred years ago. Srinivasa Ramanujan, a clerk in Bombay, mailed long screeds of mathematical formulas to G. H. Hardy, a professor at Trinity College, Cambridge. Fortunately, Hardy had the percipience to recognize that Ramanujan was not the typical green-ink scribbler who finds numerical patterns in the Bible or the pyramids and that his writings betrayed preternatural insight. Hardy arranged for Ramanujan to come to Cambridge and did all he could to foster Ramanujan's genius; sadly, culture shock and poor health brought him an early death.

The Internet enables far wider participation in front-line science; it levels the playing field between researchers in major centers and those in relative isolation, hitherto handicapped by inefficient communication. It has transformed the way science is communicated and debated. More fundamentally, it changes how research is done, what might be discovered, and how students learn.

And it allows new styles of research. For example, in the old days astronomical information, even if in principle publicly available, was stored on delicate photographic plates. These were not easily accessible, and they were tiresome to analyze. Now such data (and, likewise, large datasets in genetics or particle physics) can be accessed and downloaded anywhere. Experiments, and natural events such as tropical storms or the impact of a comet on Jupiter, can be followed in real time by anyone who's interested. And the power of huge computing networks can be deployed on large datasets.

Indeed, scientific discoveries will increasingly be made by "brute force" rather than by insight. IBM's Deep Blue beat Garry Kasparov not by thinking like him but by exploiting its speed to explore a huge variety of options. There are some high-priority scientific quests—for instance, the recipe for a room-temperature superconductor, or the identification of key steps in the origin of life—that may yield most readily neither to insight nor to experiment but to exhaustive computational searches.

Paul Ginsparg's arXiv.org archive transformed the literature of physics, establishing a new model for communication over the whole of science. Far fewer people today read traditional journals. These have so far survived as guarantors of quality. But even this role may soon be trumped by a more informal system of quality control, signaled by the approbation of discerning readers (analogous to the grading of restaurants by gastronomic critics), by blogs, or by Amazon-style reviews.

Clustering of experts in actual institutions will continue, for the same reason that high-tech expertise congregates in Silicon Valley and elsewhere. But the progress of science will be driven by ever more immersive technology where propinquity is irrelevant. Traditional universities will survive insofar as they offer mentoring and personal contact to their students. But it's less clear that there will be a future for the "mass university," where the students are offered little more than a passive role in lectures (generally of mediocre quality) with minimal feedback. Instead, the Internet will offer access to outstanding lectures—and in return will offer the star lecturers (and perhaps the best classroom teachers, too) a potentially global reach.

And it's not just students but those at the end of their careers whose lives the Internet can transformatively enhance. We oldies, as we become less mobile, will be able to immerse ourselves—right up to the final switch-off, or until we lose our wits completely—in an ever more sophisticated cyberspace that allows virtual travel and continuing engagement with the world.

Move Aside, Sex

Seth Lloyd

Quantum mechanical engineer, MIT; author, Programming the Universe

I think less. My goal is to transfer my brain's functions, bit by bit, to the Cloud.

When I do think, I'm lazier. There's no point in making the strenuous trek over to the library to find the source when you can get an expurgated electronic version on Google Books right away. And why go look up the exact theorem when you can find an approximate version on Wikipedia?

OK, you can get burned. Math being what it is, an approximate theorem is typically an untrue theorem. Over the years, I have found most statements in purely scientific reference articles on Wikipedia to be 99.44 percent correct. It's that last .56 percent that gets you. I just wasted three months and almost published an incorrect result because one clause in the Wikipedia statement of a theorem was wrong. It's a lucky thing the referee caught my error. In the meanwhile, however, I had used one of the great Internet innovations, the scientific preprint archive, to post the incorrect result on the Internet for everyone to see.

For hundreds of millions of years, sex was the most efficient method for propagating information of dubious provenance. The origins of all those snippets of junk DNA are lost in the sands of reproductive history. Move aside, sex: The World Wide Web has usurped your role. A single illegal download can disseminate more parasitic bits of information than a host of mating tsetse flies. Indeed, as I looked further afield, I found that it was not just Wikipedia that was in error: Essentially every digital statement of the clause in the theorem of interest was also incorrect. For better or worse, it appears that the only sure way to find the correct statement of a theorem is to trek to the library and find some book written by some dead mathematician—maybe even the one who proved the theorem in the first place.

In fact, the key to correctness probably lies not so much in the fact that the book was written by that mathematician as in the fact that the book was scrupulously edited by the editor of the series who had invited the mathematician to write the book. Prose, poetry, and theorems posted on the Internet are no less insightful and brilliant than their paper predecessors; they are simply less edited. Moreover, just when we need them most, the meticulously trained editors of our newspapers, journals, and publishing houses are being laid off in droves.

Life, too, has gone through periods of editorial collapse. During the Cambrian explosion, living systems discovered the evolutionary advantage of complex, multicellular forms. Like the digital organisms of today's Internet, the new Cambrian life-forms rewrote the rules of habitat after habitat, evolving rapidly in the process. Finally, however, they filled their environment to its carrying capacity; at that point, just being cool, complex, and multicellular was no longer enough to ensure survival. The sharp red pencil of natural selection came out and slashed away the gratuitous sequences of DNA.

For the moment, however, the ability of the Internet to propagate information promiscuously is largely a blessing. The preprint archives where scientific work (like my wrong paper) are posted for all to read are great levelers: Second- or third-world scientists with modems can access the unedited state of the art in a scientific field as it is produced, rather than months or years later. They, in turn, can produce and post their own unedited preprints, and so on. As long as computer memories keep doubling in capacity every year or two, those stacks of unedited information will keep doubling and doubling, too, swamping the useful and correct in a sea of extraneous bits. Eventually, the laws of physics themselves will stop this exponential explosion of memory space, and we will be forced, once more, to edit. What will happen then?

Don't ask me. By then, the full brain transfer to the Cloud should be complete. I hope not to be thinking at all.

Rivaling Gutenberg

John Tooby

Founder of evolutionary psychology; codirector, UC Santa Barbara's Center for Evolutionary Psychology

Obliterating whole lineages—diatoms and dinosaurs, corals and crustaceans, ammonites and amphibians—shock waves from the Yucatán meteor impact 65 million years ago rippled through the intricate interdependencies of the planetary ecosystem, turning blankets of life into shrouds in one incandescent geological instant. Knocking out keystone species and toppling community structures, these shifts and extinctions opened up new opportunities, inviting avian and mammalian adaptive radiations and other bursts of innovation that transformed the living world—and eventually opened the way for our placenta-suckled, unprecedentedly luxuriant brains.

What with one thing and another, now here we are: The Internet and the World Wide Web that runs on it have struck our species' informational ecology with a similarly explosive impact, their shock waves rippling through our cultural, social, economic, political, technological, scientific, and even cognitive landscapes.

To understand the nature and magnitude of what is to come, consider the effects of Gutenberg's ingenious marriage of the grape press, oil-based inks, and his method for inexpensively producing movable type. Before Gutenberg, books were scarce and expensive, requiring months or years of skilled individual effort to produce a single copy. Inevitably, they were primarily prestige goods for aristocrats and clerics, their content devoted to the narrow and largely useless status or ritual preoccupations of their owners. Slow-changing vessels bearing the distant echoes of ancient tradition, books were absent from the lives of all but a tiny fraction of humanity. Books then were travelers from the past rather than signals from the present, their cargo ignorance as often as knowledge. European awareness was parochial in the strict, original sense—limited to direct experience of the parish.

Yet a few decades after Gutenberg, there were millions of books flooding Europe, many written and owned by a new, book-created middle class, full of new knowledge, art, disputation, and exploration. Mental horizons—once linked to the physical horizon just a few miles away—surged outward.

Formerly, knowledge of all kinds had been fixed by authority and embedded in hierarchy and was by assumption and intention largely static. Yet the sharp drop in the cost of reproducing books shattered this stagnant and immobilizing mentality. Printing rained new Renaissance texts and newly recovered classical works across Europe; printing catalyzed the scientific revolution; printing put technological and commercial innovation onto an upward arc still accelerating today. Printing ignited the previously wasted intellectual potential of huge segments of the population—people who, without printing, would have died illiterate, uneducated, without voice or legacy.

Printing summoned into existence increasingly diversified bodies of new knowledge, multiplied productive divisions of labor, midwifed new professions, and greatly expanded the middle class. It threw up voluntary, new meritocratic hierarchies of knowledge and productivity to rival traditional hierarchies of force and superstition. In short, the release of printing technology into human societies brought into being a vast new ecosystem of knowledge—dense, diverse, rapidly changing, rapidly growing, and beyond the ability of any one mind to encompass or any government to control.

Over the previous millennium, heretics appeared perennially, only to be crushed. Implicitly and explicitly, beyond all question, orthodoxy defined and embodied virtue. But when, after Gutenberg, heretics such as Luther gained access to printing presses, the rapid and broad dissemination of their writings allowed dissidents to muster enough socially coordinated recruits to militarily stalemate attempts by hierarchies to suppress them. Hence, the assumption of a single orthodoxy husbanded by a single system of sanctified authority was broken, beyond all recovery.

For the same reason that Communist governments would restrict access to Marx's and Engels' original writings, the Church made it a death penalty offense (to be preceded by torture) to translate the Bible into the languages people spoke and understood. The radical change in attitude toward authority, and the revaluation of minds even at the bottom of society, can be seen in William Tyndale's defense of his plan to translate the Bible into English: "I defy the Pope, and all his laws; and if God spares my life, I will cause the boy that drives the plow to know more of the Scriptures than the Pope himself." (After his translation was printed, he was arrested, tied to the stake, and strangled.) Laymen, even plowboys, who now had access to Bibles (because they could both read and afford them) decided they could interpret sacred texts for themselves without the Church interposing itself as intermediary between book and reader. Humans being what they are, religious wars followed, in struggles to make one or another doctrine (and elite) locally supreme.

Conflicts such as the Thirty Years' War (with perhaps 10 million dead and entire territories devastated) slowly awakened Europeans to the costs of violent intolerance, and starting among dissident Protestant communities the recognized prerogatives of conscience and judgment devolved onto ever smaller units, eventually coming to rest in the individual (at least in some societies, and always disputed by rulers).

Freedom of thought and speech—where they exist—were unforeseen offspring of the printing press, and they change how we think. Political assumptions that had endured for millennia became inverted, making it thinkable that political legitimacy should arise from the sanction of the governed, rather than being a natural entitlement of rulers. And science was the most radical of printing's many offspring.

Formerly, the social validation of correct opinion had been the prerogative of local force-based hierarchies, based on tradition and intended to serve the powerful. Even disputes in natural philosophy had been settled by appeals to the textual authority of venerated ancients such as Aristotle. What alternative could there be? Yet when the unified front of religious and secular authority began to fragment, logic and evidence could come into play. What makes science distinctive is that it is the human activity in which logic and evidence (suspect, because potentially subversive of authority) are allowed to play at least some role in evaluating claims.

Galileo—arguably the founder of modern science—was threatened with torture and placed under house arrest not for his scientific beliefs but for his deeper heresies about what validates knowledge. He argued that along with Scripture, which could be misinterpreted, God had written another book, the book of nature—written in mathematics but open for all to see. Claims about the book of nature could be investigated using experiments, logic, and mathematics—a radical proposal that left no role for authority in the evaluation of (nonscriptural) truth. (Paralleling Tyndale's focus on the literate lay public, Galileo wrote almost all of his books in Italian rather than in Latin.) The Royal Society, founded two decades after Galileo's death, chose as their motto *Nullius in verba* ("On the authority of no one"), a principle strikingly at variance with the pre-Gutenberg world.

The assumptions (e.g., I should be free to think about and question anything), methods (experimentation, statistical inference, model building), and content (evolutionary biology, quantum mechanics, the computational Theory of Mind) of modern thought are unimaginably different from those held by our ancestors living before Gutenberg. All this—to simplify slightly—because of a drop in the cost of producing books.

So what is happening to us, now that the Internet has engulfed us? The Internet and its cybernetic creatures have dropped, by many more orders of magnitude, the cost in money, effort, and time of acquiring and publishing information. The knowledge (and disinformation) of the species is migrating online, a click away.

To take just first-order consequences, we see all around us transformations in the making that will rival or exceed the printing revolution—for example, heating up the chain reactions of scientific, technical, and economic innovation by pulling out the moderating rods of distance and delay. Quantity, Stalin said, has a quality all its own. The Internet also unleashes monsters from the id—our evolved mental programs are far more easily triggered by images than by propositions, a reality that jihadi Websites are exploiting in our new round of religious wars.

Our generation is living through this transformation, so although our cognitive unconscious is hidden from awareness, we can at least report on our direct experience on how our thinking has shifted before and after. I vividly remember my first day of browsing: firing link after link after link, suspended in an endless elation as I surveyed possibility after possibility for twenty hours straight—something I still feel.

Now my browsing operates out of two states of mind. The first is broad, rapid, intuitive scanning, where I feel free to click without goals, in order to maintain some kind of general scientific and cultural awareness without drowning in the endless sea. The second is a disciplined, focused exploration, where I am careful to ignore partisan pulls and ad hominem distractions, to dispense with my own sympathies or annoyance, to strip everything away except information about causation and paths to potential falsification or critical tests.

Like a good Kuhnian, I attempt to pay special attention to anomalies in my favored theories, which are easier to identify now that I can scan more broadly. More generally, it seems that the scope of my research has become both broader and deeper, because both cost less. Finally, my mind seems to be increasingly interwoven into the Internet; what I store locally in my own brain seems more and more to be metadata for the parts of my understanding that are stored on the Internet.

The Shoulders of Giants

William Calvin

Neurophysiologist; emeritus professor, University of Washington School of Medicine; author, Global Fever: How to Treat Climate Change

"The way you think" is nicely ambiguous. It could be a worldview. The way I think about climate change, for example, certainly has been changed by the access to knowledge and ideas afforded by the Internet. There is no way that I could have gotten up to speed in climate science without the Web. It has changed my view of the world and its future prospects.

But, being a physiologist, I first assumed that "the way you think" was asking about process (changing one sort of stuff into another) and how my thought process had been changed by the Internet. And as it happens, I can sketch out how that might work.

A thinking process can pop up new ideas or make surprising new connections between old thoughts. So in order to explore how the Internet changes the thinking process, consider for a moment how thought normally works.

Assembling a new combination ("associations") may be relatively easy. The problem is whether the parts hang together, whether they cohere. We get a nightly reminder of an incoherent thought process from our dreams, which are full of people, places, and occasions that don't hang together very well. Awake, an incoherent collection is what we often start with, with the mind's back office shaping it into the coherent version we finally become aware of and occasionally speak aloud. Without such intellectual constructs, there is, William James said a century ago, only "one great blooming, buzzing confusion."

To keep a half-dozen concepts from blending together like a summer drink, you need some mental structuring. In saying, "I think I saw him leave to go home," with its four verbs, you are nesting three sentences inside a fourth. We also structure plans (not just anticipation but with contingencies), play games (not just a romp but with arbitrary rules), create structured music (not just rhythm but with harmony and recursion), and employ logic (in long chains).

And atop this structured capability, we have a fascination with discovering how things hang together, as when we seek hidden patterns within seeming chaos—say, doing crossword and jigsaw puzzles, doing history, doing science, and trying to appreciate a joke. Our long train of connected thoughts is why our consciousness is so different from what came before. Structuring with quality control made it possible for us to think about the past and speculate about the future, in far more depth than if we were ruled by instinct and memory alone.

I'll use creating a novel sentence for my examples, but it's much the same for new thoughts and action plans. Quality is a matter of the degree of coherence, both within a sentence and within an enlarged context. Quality control without a supervising intelligence occurs in nature. On a millennial time scale, we see a new species evolving to better fit an ecological niche. It's a copying competition biased by the environment, making some variants reproduce better than others. On the time scale of the days to weeks after our autumn flu shot, we see the immune response shaping up a better and better antibody to fit the invading molecule. Again, this is a Darwinian copying competition improving quality. My favorite creative process, operating in milliseconds to minutes, can create a new thought that is spot-on, first time out.

All are examples of the universal Darwinian process. Though often summarized by Darwin's phrase, "natural selection," it is really a process with six essential ingredients. As far as I can tell, you need:

1. A characteristic pattern (A, the stand-in for the long form—something like a bar code) that can

2. Be copied, with
3. Occasional variations (A') or compounding, where
4. Populations of A and A' clones compete for a limited territory, their relative success biased by
5. A multifaceted environment of, say, memories and instincts under which some variants do better than others (Darwin's natural selection), and where
6. The next round of variants is primarily based on the more successful of the current generation (Darwin's inheritance principle)

Such recursion is how you bootstrap quality, why we can start with subconscious thoughts as jumbled as our nighttime dreams and still end up with a sentence of quality or a chain of logic—or anticipate the punch line of a joke.

You need a quality bootstrapping mechanism in order to figure out what to do with leftovers in the refrigerator; with successive attempts running through your head as you stand there with the door open, you can often find a "quality" scheme (that is, one that doesn't require another trip to the grocery store).

So how has the Internet's connectedness changed the Darwinian creative process? For the data-gathering stage, it affords us more variants, which others have already checked for quality. Search engine speed provides them faster, so that a number can be gathered within the time constraints of working memory—say, ten minutes. When we think we have a good enough assembly, we can do a quick search to see what others have said about near-fits to our candidate. Typically, we will be forced to conclude that our candidate isn't quite right, and further Internet searches will guide us in creating new variant formulations.

We can do all of this without the Internet, but it takes time—often much longer than the time span of working memory. To then think about the modified situation requires refreshing working memory with the old stuff. The sheer speed of checking out possibilities can minimize the need for that. Even if one is working from a library carrel, getting a PDF of an article by Wi-Fi is a lot faster than chasing around in the stacks.

I recall how envious I was when the Berkeley astronomer Rich Muller described how they worked out the comet problem for explaining the timing of mass extinctions. He said that it wasn't a good week if they couldn't kill off one or two possibilities for how comets from the Oort cloud might achieve orbits sufficient to strike the Earth. A candidate would either turn out to be physically impossible or make predictions that conflicted with observations. Nothing in brain research can possibly work that fast. It takes us decades to discover better explanations and move on. They could do it in a week.

And that's how I have been feeling about the Internet's expansion of quick access to knowledge and ideas. You can stand on the shoulders of a lot more giants at the same time.

Brain Candy and Bad Mathematics

Mark Pagel

Professor of evolutionary biology, University of Reading, United Kingdom; external professor, Santa Fe Institute

The Internet isn't changing the way I or anybody else thinks. We know this because we can still visit some people on Earth who don't have the Internet, and they think the same way we do. My general-purpose thinking circuits are hardwired into my brain from genetic instructions honed over millions of years of natural selection. True, the brain is plastic, and it responds to the way it is brought up by its user, or to the language it has been taught to speak, but its fundamental structure is not changed this way, except perhaps in extremis—maybe eight hours per day of computer games.

But the Internet does takes advantage of our appetites, and this changes our thoughts, if not the way we think. Our brains have appetites for thinking, learning, feeling, hearing, and seeing. They like to be used. It is why we do crossword puzzles and brainteasers, read books and visit art galleries, watch films, and play or listen to music. Our brain appetites act as spurs to action, in much the same way our emotions do, or much the same way our other appetites—for food and sex—do. Those of us throughout history who have acted on our world—even if just to wonder why fires start, why the wind blows out of the southwest, or what would happen if we combined heat with clay—will have been more successful than those of us who sat around waiting for things to happen.

So the Internet is brain candy to me and, I suspect, to most of us—it slakes our appetite to keep our brain occupied. That moment when a search engine pops up its 1,278,000 search results to my query is a moment of pure injection of glucose into my brain. It loves it. That's why so many of us keep going back for more. Some think that's why the Internet is going to make us lazy, less literate, and less numerate, that we will forget what lovely things books are, and so on. But even as brain candy, the Internet's influence on these sorts of capabilities and pleasures is probably not as serious as the curmudgeons and troglodytes would have you believe. They will be the same people who grumbled about the telegraph, trains, the motorcar, the wireless, and television.

There are far more interesting ways that the Internet changes our thoughts, and especially the conclusions we draw, and it does this also by acting on our appetites. I speak of contagion, false beliefs, neuroses (especially medical and psychological neuroses), conspiracy theories, and narcissism. The technical point is this: The Internet tricks us into doing bad mathematics; it gets us to do a mathematical integration inside our brains that we don't know how to do. What? In mathematics, integration is a way of summing an infinite number of things. It is used to calculate quantities like volumes, areas, rates, and averages. Our brains evolved to judge risks, to assess likelihood or probabilities, to defend our minds against undue worry, and to infer what others are thinking, by sampling and summing or averaging across small groups of people, most probably the people in my tribe. They do this automatically, normally without us even knowing about it.

In the past, my assessment of the risk of being blown up by a terrorist, or of getting swine flu, or of my child being snatched by a pedophile on the way to school was calculated from the steady input of information I would have received mainly from my small local group—because these were the people I spoke to or heard from, and these were the people whose actions affected me.

What the Internet does, and what mass communication does more generally, is to sample those inputs from the 6.8 billion people on Earth. But my brain is still considering that these inputs have arisen from my local community, because that is the case its assessment circuits were built for. That is what I mean by "bad mathematics." My brain assumes a small denominator (that is, the bottom number in a fraction), and therefore the answer to the question of how likely something is to happen is too big.

So when I hear every day of children being snatched, my brain gives me the wrong answer to the question of risk: It has divided a big number (the children snatched all over the world) by a small number (the tribe). Call this the "Madeleine McCann effect." We all heard months of coverage of this sad case of kidnapping—as of this writing, still unresolved—and it has caused us undue worry, although trivial compared with what the McCanns suffered. The effects of the bad mathematics don't stop with judging risks. Doing the integration wrong means that contagion can leap across the Internet. Contagion is a form of risk assessment with an acutely worrying conclusion. Once it starts on the Internet, everyone's bad mathematics make it explode. So do conspiracy theories: If it seems that everyone is talking about something, it must be true! But this is just the wrong denominator again. Neuroses and false beliefs are buttressed. We all worry about our health; in the past, we would look around us and find that no one else was worrying or ill. But consult the Internet and 1,278,000 people (at least!) are worrying, and they've even developed Websites to talk about their worry. The 2009 swine flu pandemic has been a damp squib, but you wouldn't have known that from the frenzy.

The bad mathematics can also give us a sense that we have something useful to say. We'd all like to be taken seriously, and evolution has probably equipped us to think we are more effective than we really are; it seeds us with just that little bit of narcissism. A false belief, perhaps, but better for evolution to err on the side of getting us to believe in ourselves than not to. So we go on the Internet and make Websites, create Facebook pages, contribute to YouTube, and write blogs, and—surprise!—it appears that everyone is reading them, because look at how many people are leaving comments! Another case of the wrong denominator.

The maddening side of all this is that neither I nor most others can convince ourselves to ignore these worries, neuroses, narcissistic beliefs, and poor assessments of risk—to ignore our wrong thoughts—precisely because the Internet has not changed the way we think.

Publications Can Perish

Robert Shapiro

Professor emeritus of chemistry and senior research scientist, New York University; author, Planetary Dreams: The Quest to Discover Life Beyond Earth

The Internet has made it far easier for professionals to access and search the scientific literature. Unfortunately, it has also increased the chances that we will lose part or all of that literature.

When I was young, I imagined that everything I wrote would be preserved forever. Future biographers would seek out every letter, diary, and memorandum to capture the essence of my creativity. My first laboratory notebook still reflected the same emotions. On page 1, I had printed, very legibly, the following preface: "To Posterity: This volume contains the authentic record of ingenious and original chemical research conducted by Robert Shapiro, currently a graduate student of organic chemistry at Harvard University."

Reality gradually whittled down my grandiosity, and I recognized that my published papers had the best chance of survival. The New York University library carried bound journals that dated from the nineteenth century, and the articles they contained had obviously outlived their authors. As the number of my own published works grew, curiosity prompted me to select one of them and track its effect. I deliberately picked one of minor importance.

A generation ago, a persistent PhD student and I had failed in an effort to synthesize a new substance of theoretical interest. We had, however, prepared some other new compounds and improved some methods, so I wrote a paper that was published in 1969 in the *Journal of Organic Chemistry*. Had our results ever mattered to anyone? Using new computer-driven search tools, I could quickly check whether it had ever been noticed. To my surprise, I found that eleven papers and some patents had cited our publication, up to 2002. In one instance, our work provided a starting point for the preparation of new tranquilizers. I imagined that in the distant future other workers might pull the appropriate volume off a library shelf and find my work to be some help. I did not foresee that such bound volumes might no longer exist.

The Journal of Organic Chemistry started in 1936 and continues to the present. Its demands on library shelf space have increased over time: The first volume contained only 583 pages, whereas the 2009 edition had 9,680. The arrival of the Internet rescued libraries from the space crisis created by the proliferation of new journals and the vast increase in the size of existing ones. Many paper subscriptions were replaced by electronic ones, and past holdings were converted to digital form. It is not hard to imagine a future time when paper copies of the scientific literature will no longer exist. Many new journals are appearing only in digital form.

This conversion has produced many benefits for readers. In the past, I had to leave my office, ride an elevator, walk several blocks, take another elevator, and make my way through a maze of shelves to find a paper I needed. Occasionally the issue I wanted was being used by someone else or had been misplaced and I had traveled in vain. Now I can bring most papers that I want onto a computer screen in my office or at home in a matter of minutes. I can store the publication in my computer or print out a copy if I wish. But with this gain in the accessibility of the literature of science has come an increase in its vulnerability.

Materials that exist in one or a few copies are inherently at greater risk than those that are widely distributed. A Picasso painting might be destroyed, but the Bible will survive. Alexander Stille, in *The Future of the Past*, reported that the works of Homer and Virgil survived from antiquity because their great popularity led them to be copied and recopied. On the other hand, only 9 of Sophocles' 120 plays have survived. Before the Internet, I could take pride that each of my papers was present in hundreds or thousands of libraries across the globe. Their survival into the future seemed assured by the protection afforded by multiple copies. The same applied, of course, to the rest of the scientific literature.

Thousands of paper copies of the literature have now been replaced by a few electronic records stored in computers. Furthermore, the storage medium is fragile. Some paper manuscripts have survived for centuries. The lifetimes of the various disks, drives, and tapes currently used for digital storage are unknown but are commonly estimated in decades. In some cases, works available only in electronic form have disappeared much more rapidly for another reason—lack of maintenance of the sites. One survey found that 12 percent of the Internet addresses cited in three prestigious medical and scientific journals were extinct two years after publication.

Such difficulties are unlikely to affect such prestigious sources as *The Journal of Organic Chemistry*. But material stored only on the Internet is far more vulnerable to destruction than the same material present in multiple paper copies. Electrical breakdown can disrupt access for a time, while cyberterrorism, civic disturbances, war, and a variety of natural catastrophes could destroy part or all of the storage system, leading to the irretrievable loss of sections of the scientific literature. Anton Zeilinger wrote in answer to a previous *Edge* question that a nuclear explosion outside Earth's atmosphere would cause all computers, and ultimately society, to break down.

How has this changed my thinking? I no longer write with the expectation of immortality in print. I am much more tempted to contribute to Internet discussion forums, blogs, and media that may not persist. I seek my reward from the immediate response my efforts may bring, with little thought to the possibility that some stranger may see my words centuries from now and wonder about the life that was led by the person who wrote them.

Will the Great Leveler Destroy Diversity of Thought?

Frank J. Tipler

Professor of mathematical physics, Tulane University; author, The Physics of Immortality

The Internet first appeared long after I had received my PhD in physics, and I was slow to use it. I was trained in library search techniques: look up the subject in *Science Abstracts* (a journal now made defunct by the Internet), then go to the archived full article in the journal shelved nearby. Now I simply search the topics in the Science Citation Index (SCI) and then go to the journal article, available online.

These Internet versions of journals and abstracts have one disadvantage at present: My university can afford only a limited window for the search. I can use the SCI only back ten years; moreover, most e-journals have not yet converted their older volumes to online format (nor can my university often afford to pay for access to these older print journals).

So the Internet causes scientific knowledge to become obsolete faster than was the case with the older print media. A scientist trained in the print media tradition is aware that there is knowledge stored in the print journals, but I wonder if the new generation of scientists, who grew up with the Internet, is aware of this. Print journals were forever. They may have merely gathered dust for decades, but they could still be read by later generations. I can no longer read my own articles stored on the floppy disks of the 1980s, because computer technology has changed too much. Will information stored on the Internet become unreadable to later generations because of data storage changes—and will the knowledge thus be lost?

At the moment, the data are accessible. More important, raw experimental data are becoming available to theorists, like myself, via the Internet. It is well known from the history of science that experimentalists quite often do not appreciate the full significance of their own observations. "A new phenomenon is first seen by someone who did not discover it" is one way of expressing this fact. Now that the Internet allows the experimenter to post her data, we theorists can individually analyze them.

Let me give an example from my own work. Standard quantum mechanics asserts that an interference pattern of electrons passing through a double slit must have a certain distribution as the number of electrons approaches infinity. However, this same standard quantum mechanics does not give an exact description of the rate at which the final distribution will be approached. Many-worlds quantum mechanics, in contrast, gives us a precise formula for this rate of approach, since according to the many-worlds interpretation, physical reality is not probabilistic at all but more deterministic than the universe of classical mechanics. (According to many-worlds quantum mechanics, the wave function measures the density of worlds in the multiverse, rather than a probability.)

Experimenters—indeed, undergraduate students in physics—have observed the approach to the final distribution, but they have never tried to compare their observations with any rate-of-approach formula, since, according to standard quantum mechanics, there is no rate-of-approach formula. Using the Internet, I was able to find raw data on electron interference, which I used to test the many-worlds formula. Most theorists can tell a similar story.

But I sometimes wonder if later generations of theorists will be able to tell such a story. Discoveries can be made by analyzing raw data posted online today, but will this always be true? The great physicist Richard Feynman often claimed that "there will be no more great physicists." Feynman believed that the great physicists were those who looked at reality from a point of view different from that of other scientists. He argued, in *Surely You're Joking, Mr. Feynman*, that all of his own achievements were due not to an IQ higher than other physicists' but to his having a "different bag of tricks." Feynman thought future generations of physicists would all have the same bag of tricks and consequently would be unable to move beyond the consensus view. Everyone would think the same way.

The Internet is currently the great leveler: It allows everyone access to exactly the same information. Will this ultimately destroy diversity of thought? Or will the tendency of people to form isolated groups on the Internet preserve that all-important diversity so that although all scientists have equal access in principle, there are still those who will look at the raw data in a different way from the consensus?

We Have Become Hunter-Gatherers of Images and Information

Lee Smolin

Physicist, Perimeter Institute; author, The Trouble with Physics

The Internet hasn't, so far, changed how we think. But it has radically altered the contexts in which we think and work.

The Internet offers a vast realm for distraction, but then so does reading and television. The Internet is an improvement on television in the same way that Jane Jacobs's bustling neighborhood sidewalk is an improvement on the dullness of suburbia. The Internet requires an active engagement, and as a result it is full of surprises. You don't watch the Internet; you search and link. What is important for thought about the Internet is not the content, it is the new activity of being a searcher, with the world's store of knowledge and images at your fingertips.

The miracle of the browser is that it can show you any image or text from that storehouse. We used to cultivate thought; now we have become hunter-gatherers of images and information. This speeds things up a lot, but it doesn't replace the hard work in the laboratory or notebook that prepares the mind for a flash of insight. Nonetheless, it changes the social situation of that mind. Scholars used to be more tied to the past, through texts in libraries, than to their contemporaries. The Internet reverses that, by making each of our minds a node in a continually evolving network of other minds.

The Internet is also itself a metaphor for the emerging paradigm of thought in which systems are conceived as networks of relationships. To the extent that a Web page can be defined only by what links to it and what it links to, it is analogous to one of Leibniz's monads. But Web pages have content and so are not purely relational. Imagine a virtual world abstracted from the Internet by deleting the content so that all that remained was the links. This is an image of the universe according to relational theories of space and time; it is also an image of the neural network in the brain. The content corresponds to what is missing in those models; it corresponds to what physicists and computer scientists have yet to understand about the difference between a mathematical model and an animated world or conscious mind.

Perhaps when the Internet has been soldered into our glasses or teeth, with the screen replaced by a laser making images directly on our retinas, there will be deeper changes. But even in its present form, the Internet has transformed how we scientists work.

The Internet flattens communities of thought. Blogs, e-mail, and Internet databases put everyone in the community on the same footing. There is a premium on articulateness. You don't need a secretary to maintain a large and varied correspondence.

Since 1992, research papers in physics have been posted on an Internet archive, arXiv.org, which has a daily distribution of just-posted papers and complete search and cross-reference capabilities. It is moderated rather then refereed, and the refereed journals now play no role in spreading information. This gives a feeling of engagement and responsibility: Once you are a registered member of the community, you don't have to ask anyone's permission to publish your scientific results.

The Internet delocalizes your community. You participate from wherever you are. You don't need to travel to listen to or give talks and there is less reason to go into the office. A disinclination to travel is no reason not to stay by current reading the latest papers and blog postings.

It used to be that physics preprints were distributed by bulk mail among major research institutes, and there was thus a big advantage to being at a major university in the United States; everyone else was working with the handicap of being weeks to months behind. The increasing numbers and influence of scientists working in Asia and Latin America and the dominance of European science in some fields is a consequence of the Internet.

The Internet synchronizes the thinking of global scientific communities. Everyone gets the news about the new papers at the same time. Gossip spreads just as fast, on blogs. Announcements of new experimental results are videocast through the Internet as they happen.

The Internet also broadens communities of thought. Obscure thinkers to whom you had to be introduced and who published highly original work sporadically and in hard-to-find places now have Web pages and post their papers alongside everyone else's. And it creates communities of diverse thinkers who otherwise would not have met, like the community we celebrate every year at this time when we answer the *Edge* annual question.

The Human Texture of Information

Jon Kleinberg

Professor of computer science, Cornell University; coauthor (with David Easley), Networks, Crowds, and Markets: Reasoning About a Highly Connected World

When Rio de Janeiro was announced as the site of the 2016 Summer Olympics, I was on the phone with colleagues, talking about ideas for how to track breaking news on the Internet. Curious to see how reactions to the announcement were playing out, we went onto the Web to take a look, pushing our way like tourists into the midst of a celebration that was already well under way. The sense that we were surrounded by crowds was not entirely in our imaginations: More than a thousand tweets per minute about Rio were appearing on Twitter, Wikipedians were posting continuous updates to the "2016 Summer Olympics" page, and political blogs were filled with active conversations about the lobbying of world leaders on behalf of different cities.

This is the shape that current events take online, and there is something more going on here than simple volume. Until recently, information about an event such as this would have been disseminated according to a top-down structure, consisting of an editorially assembled sampling of summaries of the official announcement, reports of selected reactions, and stories of crowds gathering at the scene. But now the information emerges from the bottom up, converging in tiny pieces from all directions. The crowd itself speaks, in a million distinct voices—a deluge of different perspectives.

The Web hasn't always looked this way. When I first used an Internet search engine in the early 1990s, I imagined myself dipping into a vast, universal library—a museum vault filled with accumulated knowledge. The fact that I shared this museum vault with other visitors was something I knew but could not directly perceive; we had the tools to engage with the information but not with one another, and so we all passed invisibly by one another.

When I go online today, all those rooms and hallways are teeming and I can see it. What strikes me is the human texture of the information—the visible conversations, the spikes and bursts of text, the controlled graffiti of tagging and commenting. I've come to appreciate the way the event and the crowd live in symbiosis, each dependent on the other—the people all talking at once about the event, but the event fully comprehensible only as the sum total of the human reaction to it. The construction feels literary in its complexity—a scene described by an omniscient narrator, jumping between different points of view, except that here all these voices belong to real, living beings and there's no master narrative coordinating them. The cacophony might make sense or it might not.

But the complexity does not arise just from all the human voices—it is accentuated by the fact that the online world is one where human beings and computational creations commingle. You bump into these computational artifacts like strange characters in a Carrollian Wonderland. There is the giant creature who has memorized everything ever written and will repeat excerpts back to you (mainly out of context) in response to your questions. There are the diaphanous forms, barely visible at the right-hand edge of your field of vision, who listen mutely as you cancel meetings and talk about staying home in bed and then mysteriously begin slipping you ads for cough medicine and pain relievers. And even more exotic characters are on the way; a whole industry works tirelessly to develop them.

The ads for cough medicine are important, and not just because they're part of what pays for the whole operation. They should continually remind you that you're part of the giant crowd as well—that everything you do is feeding into a global conversation that is not only visible but recorded. I try to reflect on what behavioral targeting algorithms must think of me—what the mosaic of my actions must look like when everything is taken into account, and which pieces of that mosaic would have been better left off the table.

The complexity of the online world means that when I use the Internet today, even for the most mundane of purposes, I find myself drawing on skills I first learned in doing research: evaluating many different observations and interpretations of the same events; asking how people's underlying perspectives, tools, and ways of behaving have shaped their interpretations; and reflecting on my own decisions as part of this process. Think about the cognitive demands this activity involves. Once the domain of scholarship, it is now something the Internet requires from us on a daily basis. It suggests that in addition to "computer literacy"—an old pursuit wherein we teach novices how to use computing technology in a purely operational sense—we need to be conveying the much more complex skill of "information literacy" to the very young: how to reason about the swirl of perspectives you find when you consume information online, how to understand and harness the computational forces that shape this information, how to reason about the subtle consequences of your own actions on the Internet.

Finally, the Internet has changed how I think professionally, as a computer scientist. In the thirteen years since I finished graduate school, the Internet has steadily and incontrovertibly advanced the argument that computer science is not just about technology but about human beings as well—about the power of human beings to collectively create knowledge and engage in self-expression on a global scale. This has been a thrilling development, and one that points to a new phase in our understanding of what people and technology can accomplish together, and about the world we've grown to jointly inhabit.

Not at All

Steven Pinker

Johnstone Family Professor, Department of Psychology; Harvard University; author, The Stuff of Thought: Language as a Window into Human Nature

As someone who believes both in human nature and in timeless standards of logic and evidence, I'm skeptical of the common claim that the Internet is changing the way we think. Electronic media aren't going to revamp the brain's mechanisms of information processing, nor will they supersede *modus ponens* or Bayes's theorem. Claims that the Internet is changing human thought are propelled by a number of forces: the pressure on pundits to announce that this or that "changes everything"; a superficial conception of what "thinking" is that conflates content with process; the neophobic mindset that "if young people do something that I don't do, the culture is declining." But I don't think the claims stand up to scrutiny.

Has a generation of texters, surfers, and Twitterers evolved the enviable ability to process multiple streams of novel information in parallel? Most cognitive psychologists doubt it, and recent studies by Clifford Nass of Stanford University's Communication Department confirm their skepticism. So-called multitaskers are like Woody Allen after he took a speed-reading course and devoured *War and Peace* in an evening. His summary: "It was about some Russians."

Also widely rumored are the students who cannot write a paper without instant-message abbreviations, emoticons, and dubious Web citations. But students indulge in such laziness only to the extent that their teachers let them get away with it. I have never seen a paper of this kind, and a survey of student papers by Stanford English professor Andrea Lunsford shows that they are mostly figments of the pundits' imaginations.

The way that intellectual standards constrain intellectual products is nowhere more evident than in science. Scientists are voracious users of the Internet and other computer-based technologies that are supposedly making us stupid, such as PowerPoint, electronic publishing, and e-mail. Yet it would be ludicrous to suggest that scientists think differently than they did a decade ago or that the progress of science has slowed.

The most interesting trend in the development of the Internet is not how it is changing people's ways of thinking but how it is adapting to the way people think. The leap in Internet usage that accompanied the appearance of the World Wide Web more than a decade ago came from its user interface, the graphical browser, which worked around the serial, line-based processing of the actual computer hardware to simulate a familiar visual world of windows, icons, and buttons. The changes we are seeing more recently include even more natural interfaces (speech, language, manual manipulation), better emulation of human expertise (as in movie, book, or music recommendations, and more intelligent search), and the application of Web technologies to social and emotional purposes (such as social networking, and sharing of pictures, music, and video) rather than just the traditional nerdy ones.

To be sure, many aspects of the life of the mind have been affected by the Internet. Our physical folders, mailboxes, bookshelves, spreadsheets, documents, media players, and so on have been replaced by software equivalents, which has altered our time budgets in countless ways. But to call it an alteration of "how we think" is, I think, an exaggeration.

This Is Your Brain on Internet

Terrence Sejnowski

Computational neuroscientist, Salk Institute; coauthor (with Patricia Churchland), The Computational Brain

What is the impact of spending hours each day in front of a monitor, surfing the Internet and playing games? Brains are highly adaptable, and experiences have long-term effects on its structure and function. You are aware of some of the changes and call it your memory, but this is just the tip of the iceberg. We are not aware of more subtle changes, which nonetheless can affect your perception and behavior. These changes occur at all levels of your brain, from the earliest perceptual levels to the highest cognitive levels.

Priming is a dramatic example of unconscious learning, in which a brief exposure to an image or a word can affect how you respond to the same image or word, even in degraded forms, many months later. In one experiment, subjects briefly viewed the outlines of animals and other familiar objects, and seventeen years later they could identify the animals and objects, above chance levels, from versions in which half the outlines had been erased. Some of these subjects did not even remember participating in the original experiment. With conceptual priming, an object like a table can prime the response to a chair. Interestingly, priming decreases reaction times and is accompanied by a decrease in brain activity—the brain becomes faster and more efficient.

Brains, especially youthful ones, have an omnivorous appetite for information, novelty, and social interaction, but it is less obvious why we are so good at unconscious learning. One advantage of unconscious learning is that it allows the brain to build up an internal representation of the statistical structure of the world: the frequency of neighboring letters in words, say, or the textures, forms, and colors that make up images. Brains are also adept at adapting to sensorimotor interfaces. We first adapted to clunky keyboards, then to virtual pointers to virtual files, and now to texting with fingers and thumbs. As you become an expert at using it, the Internet, as with other tools, becomes an extension of your brain.

Are the changes occurring in your brain as you interact with the Internet good or bad for you? Adapting to the touch and feel of the Internet makes it easier for you to extract information, but a better question is whether the changes in your brain will improve your fitness. There was a time, not long ago, when CEOs didn't use the Internet because they had never learned to type—but these folks are going extinct and have been replaced with more Internet-savvy managers.

Gaining knowledge and skills should benefit your survival, but not if you spend *all* your time immersed in the Internet. The intermittent rewards can become addictive, hijacking your dopamine neurons (which predict future rewards). But the Internet has not been around long enough—and is changing too rapidly—for us to know what the long-term effects will be on brain function. What is the ultimate price for omniscience?

The Sculpting of Human Thought

Donald Hoffman

Cognitive scientist, UC Irvine; author, Visual Intelligence: How We Create What We See

Human thought has many sculptors, and each wields special tools for distinct effects. Is the Internet in the tool kit? That depends on the sculptor.

Natural selection sculpts human thought across generations and at geologic time scales. Fitness is its tool, and human nature, our shared endowment as members of a species, is among its key effects. Although the thought life of each person is unique, one can discern patterns of thought that transcend racial, cultural, and occupational differences; similarly, although the face of each person is unique, one can discern patterns of physiognomy—two eyes above a nose above a mouth—that transcend individual differences.

Is the Internet in the tool kit of natural selection? That is, does the Internet alter our fitness as a species? Does it change how likely we are to survive and reproduce? Debate on this question is in order, but the burden is surely on those who argue no. Our inventions in the past have altered our fitness: arrowheads, agriculture, the control of fire. The Internet has likely done the same.

But has the Internet changed the patterns of thought that transcend individual differences? Not yet. Natural selection acts over generations; the Internet is but one generation old. The Internet is in the tool kit but has not yet been applied. Over time, as the Internet rewards certain cognitive skills and ignores or discourages others, it could profoundly alter even the basic patterns of thought we share as a species. The catch, however, is "over time." The Internet will evolve new offspring more quickly than *Homo sapiens*, and they, rather than the Internet, will alter human nature. These offspring will probably no more resemble the Internet than *Homo sapiens* resembles amoebae.

Learning sculpts human thought across the lifetime of an individual. Experience is its tool, and unique patterns of cognition, emotion, and physiology are its key effects. Psychologists Marcel Just and Timothy Keller found that poor readers in elementary school could dramatically improve their skills with six months of intensive training and that white-matter connections in the left hemispheres of their brains increased measurably in the process.*

There are, of course, endogenous limits to what can be learned, and these limits are largely a consequence of mutation and natural selection. A normal infant exposed to English will learn to speak English, but the same infant exposed to C++ or HTML will learn little.

Is the Internet in the tool kit of learning? No doubt. Within the endogenous limits of learning set by your genetic inheritance, exposure to the Internet can alter how you think no less than can exposure to language, literature, or mathematics. But the endogenous limits are critical. Multitasking, for instance, might be a useful skill for exploiting in parallel the varied resources of the Internet, but genuine multitasking, at present, probably exceeds the limitations of the attentional system of *Homo sapiens.* Over generations, this limitation might ease. What the Internet cannot accomplish as a tool of learning it might eventually accomplish as a tool of natural selection.

Epigenetics (the study of changes in appearance or gene expression caused by mechanisms other than changes in DNA) sculpts human thought within a lifetime and across a few generations. Experience and environment are its guides, and shifts in gene expression triggering shifts in cognition, emotion, and physiology are its relevant effects. Neuroscientist Timothy Oberlander and colleagues found that

---

* T. A. Keller and M. A. Just, "Altering Cortical Connectivity: Remediation-Induced Changes in the White Matter of Poor Readers," *Neuron* 64 (2009): 624–31.

a mother's depression can change the expression of the NR3C1 gene in her newborn, leading to the infant's increased reactivity to stress.* Childhood abuse similarly can lead to persistent feelings of anxiety and acute stress in a child, fundamentally altering its thought life.

Is the Internet in the tool kit of epigenetics? Possibly, but no one knows. The field of epigenetics is young, and even the basic mechanisms by which transgenerational epigenetic effects are inherited are not well understood. But the finding that parental behavior can alter gene expression and thought life in a child certainly leaves open the possibility that other behavioral environments, including the Internet, can do the same.

Thus, in sum, the relevance of the Internet to human thought depends on whether one evaluates this relevance phylogenetically, ontogenetically, or epigenetically. Debate on this issue can be clarified by specifying the framework of evaluation.

* T. F. Oberlander et al., "Prenatal Exposure to Maternal Depression, Neonatal Methylation of Human Glucocorticoid Receptor Gene (NR3C1) and Infant Cortisol Stress Responses," *Epigenetics* 3, 2 (2008): 97–106.

What Kind of a Dumb Question Is That?

Andy Clark

Philosopher and cognitive scientist, University of Edinburgh; author, Supersizing the Mind: Embodiment, Action, and Cognitive Extension

How is the Internet changing the way I think? There is something tremendously slippery—but actually, despite my attention-seeking title, interestingly and importantly slippery—about this question. To see what it is, reflect first that the question has an apparently trivial variant: "Is the Internet changing the things you think?"

This is a question that has all kinds of apparently shallow answers. The Internet is certainly changing *what* I think (it makes all kinds of information and views available to me that would not be otherwise). The Internet is also changing when I think it, how long it takes me to think it, and what I do with it when I've finished thinking it. The Internet is even changing how I carry out lots of the thinking, making that a rather more communal enterprise than it used to be (at least in my area, which is scientifically informed philosophy of mind).

But that all sounds kind of shallow. We all know the Internet does that. What the question means to get at, surely, is something slippery but deeper, something that may or may not be true, viz.: "Is the Internet changing the nature of your thinking?"

It's this question, I suggest, that divides the bulk of the respondents. There are those who think the nature of human thinking hasn't altered at all and those who think it is becoming radically transformed. The question I want to ask in return, however, is simply this: "How can we know?" I don't think this question has any easy answer.

One place to start might be to distinguish what we think from the routines we employ to think it. By "routines" I mean something in the ballpark of an algorithm—some kind of computational recipe for solving a problem or class of problems. Once we make this distinction, it can seem (but this may turn out to be a deep illusion) plain sailing. For it then seems the question is simply one for science to figure out. For how would you know whether the way you were thinking had been altered? If what you tend to think alters, does that imply that the way you are thinking it must be altered, too? I guess not. Or try it the other way around. If what you tend to think and believe remains the same, does that imply that the way you're thinking it remains the same? I guess not.

The most we can tell from our armchairs, it seems to me, is that what we're thinking (and when we tend to think it) is in some way altering. But of course, there can be no doubt that the Internet alters what we tend to think and when. If it didn't, we wouldn't need it. So that's true but kind of trivial.

Otherwise put: From my philosopher's armchair, all I know is what anyone else knows, and that's all about content. I know (on a good day) what I think. But as to the routines I use to think it, I have as little idea as I have (from my armchair) of what moves the planets. I have access to the results, not the means. Insofar as I have any ideas at all about what routines or means I use to do my thinking, those ideas are no doubt ragingly false. At best, they reflect how I think I think my thoughts, not how I do.

So far, so good. At this point, it looks as if we must indeed turn to some kind of experimental science to find the answer to any nontrivial reading of the question.

Is the Internet changing the way I think? Let's put on our lab coats and go find out.

But how?

Suppose we go looking for some serious neural changes in heavy Internet users. Problem: There are bound to be some changes, as surfing the Web is a skill and skills alter brains. But when does some such change count as a change to the way we think? Does learning to play the piano change the way I think? Presumably not, in the kind of way that the question means. Even quite large neural changes might not effect a change in the way we think. Perhaps it's just the same old way being employed to do some new stuff. Conversely, even a quite small neural change might amount to the installation of a whole new computational architecture (think of adding a recurrent loop to a simple neural network—a small neural change with staggeringly profound computational consequences).

It gets worse.

Not only is it unclear what science needs to discover, it is unclear where science ought to look to discover (or not discover) it.

Suppose we convince ourselves, by whatever means, that as far as the basic mode of operation of the brain goes, Internet experience is not altering it one whit. That supports a negative answer only if we assume that the routines that fix the "nature of human thinking" must be thoroughly biological—that they must be routines running within, and only within, the individual human brain. But surely it is this assumption that our experiences with the Internet (and with other "intelligence amplifiers" before it) most clearly call into question. Perhaps the Internet is changing "the way you think" by changing the circuits that get to implement some aspects of human thinking, providing some hybrid (biological and nonbiological) circuitry for thought itself. This would be a vision of the Internet as a kind of worldwide supracortex. Since this electronic supracortex patently does not work according to the same routines as, say, the neocortex, an affirmative answer to our target question seems easily in the cards.

But wait. Why look there in the first place? What exactly determines (or, better, what should determine) where we look for the circuitry whose operational profile, even assuming we can find it, determines the "way we think"?

This is a really hard question—and, sad to say, I don't know how to answer it. It threatens to bring us all the way back to where we started, with content. For perhaps one way to motivate an answer is to look for deep and systematic variation in human performances in various spheres of thought. But even if we find such variation, those who think that our "ways of thinking" remain fundamentally unaltered can hold their ground by stressing that the basic mode of neural operation is unaltered and has remained the same for (at least) tens of thousands of years.

Deep down, I suspect that our two interrogative options—the trivial-sounding question about what we think and the deep-sounding one about the nature of our thinking—are simply not as distinct as the fans of either response (Yes, the Internet is changing the way we think/No, it isn't) might wish.

But I don't know how to prove this.

Dammit.

Public Dreaming

Thomas Metzinger

Philosopher; director of the Theoretical Philosophy Group at the Department of Philosophy of the Johannes Gutenberg–Universität Mainz; author, The Ego Tunnel

I heard a strange, melodic sound from the left and turned away from the Green Woman. As I shifted my gaze toward the empty landscape, I noticed that something wasn't quite right. What I saw, the hills and the trees, were as real as could be—but somehow they hadn't come into view as they would in real life. Somehow it wasn't quite in real time. There was a slightly different temporal dynamics to the way the scene popped up, an almost unnoticeable delay, as if I were surfing the Web, clicking my way onto another page. But I wasn't surfing. I had just talked to the Green Woman—and no, my right index finger wasn't clicking and my right hand wasn't lying on a mouse pad; it hung by my side, completely relaxed, as I gazed at the empty landscape of hills and trees. In a flash of excitement and disbelief, it dawned on me: I was dreaming!

I have always been interested in lucid dreams and have written about them extensively. They interest consciousness researchers because you can go for a walk through the dynamics of your own neural correlate of consciousness, unconstrained by external input, and look at the way the experience unfolds from the inside. They are interesting to philosophers, too. You can ask the dream characters you encounter what they think about notions such as "virtual embodiment" and "virtual selfhood" and whether they believe they have a mind of their own. Unfortunately, I have lucid dreams only rarely—once or twice a year. The episode just recited was the beginning of my last one, and a lot of things dawned on me at once besides the fact that I was actually inside my own head. The Internet is reconfiguring my brain, not just changing the way I think. It already penetrates my dream life.

Sure, for academics the Internet is a fantastic resource—almost all the literature at your fingertips, wonderfully efficient ways of communicating and collaborating with researchers around the world, an endless source of learning and inspiration. Something that leads you right into attention deficit disorder. Something that gets you hooked. Something that is changing us in our deepest core.

But it's about much more than cognitive style alone. For those of us intensively working with it, the Internet has become a part of our self-model. We use it for external memory storage, as a cognitive prosthesis, and for emotional autoregulation. We think with the help of the Internet, and it helps us determine our desires and goals. Affordances infect us, subtly eroding the sense of control. We are learning to multitask, our attention span is becoming shorter, and many of our social relationships are taking on a strangely disembodied character. Some software tells us, "You are now friends with Peter Smith!" when we were just too shy to click the Ignore button.

"Online addiction" has long been a technical term in psychiatry. Many young people (including an increasing number of university students) suffer from attention deficits and can no longer focus on old-fashioned, serial symbolic information; they suddenly have difficulty reading ordinary books. Everybody has heard about midlife burnout and rising levels of anxiety in large parts of the population. Acceleration is everywhere.

The core of the problem is not cognitive style but attention management. The ability to attend to our environment, our feelings, and the feelings of others is a naturally evolved feature of the human brain. Attention is a finite commodity, and it is absolutely essential to living a good life. We need attention in order to truly listen to others—and even to ourselves. We need attention to truly enjoy sensory pleasures, as well as for efficient learning. We need it in order to be truly present during sex, or to be in love, or when we are just contemplating nature. Our brains can generate only a limited amount of this precious resource every day. Today the advertisement and entertainment industries are attacking the very foundations of our capacity for experience, drawing us into a vast and confusing media jungle, robbing us of our scarce resource in ever more persistent and intelligent ways. We know all that, but here's something we are just beginning to understand: The Internet affects our sense of selfhood, and it does so on a deep functional level.

Consciousness is the space of attentional agency: Conscious information is exactly that information in your brain to which you can deliberately direct your attention. As an attentional agent, you can initiate a shift in attention and, as it were, direct your inner flashlight at certain targets: a perceptual object, say, or a specific feeling. In many situations, people lose the property of attentional agency, and consequently their sense of self is weakened. Infants cannot control their visual attention; their gaze seems to wander aimlessly from one object to another, because this part of their ego is not yet consolidated. Another example of consciousness without attentional control is the nonlucid dream state. In other cases, too, such as severe drunkenness or senile dementia, you may lose the ability to direct your attention—and, correspondingly, feel that your "self" is falling apart.

If it is true that the experience of controlling and sustaining your focus of attention is one of the deeper layers of phenomenal selfhood, then what we are currently witnessing is not only an organized attack on the space of consciousness per se but also a mild form of depersonalization. New medial environments may therefore create a new form of waking consciousness that resembles weakly subjective states—a mixture of dreaming, dementia, intoxication, and infantilization. Now we all do this together, every day. I call it public dreaming.

The Age of (Quantum) Information?

Anton Zeilinger

Physicist, University of Vienna; scientific director, Institute of Quantum Optics and Quantum Information, Austrian Academy of Sciences; author, Dance of the Photons: From Einstein to Quantum Teleportation

Yes, I have learned, like many others,

> To write short e-mails, because people don't want to read beyond line ten
>
> To write single-issue e-mails, because second or third issues get lost
>
> To check my e-mails on the iPhone or BlackBerry every five minutes, because the important message could arrive at any moment
>
> To expect that our brain function will significantly be reduced in the coming decades to very simple decision making

And so on and so on.

Well, seriously, I find it utterly impressive how the notion of information is becoming more and more important in our society. Or, rather, the notion of what we think information is. What is information? From a pragmatic, operational point of view, one could argue that information is the truth value of a proposition. Is it raining now? Yes/no. Do airplanes fly because they are lighter than air? Yes/no. Does she love me? Yes/no.

Evidently, there are questions that are easy to answer and others that are difficult or maybe even impossible to answer in a reliable way—such as the last one. Whereas for the first two questions we can devise scientific procedures for how to decide them (even including borderline cases), for the last question such an algorithm seems impossible, even though some of our biology friends try to convince us that it is just a matter of deterministic procedures in our brains and our bodies. There are other questions that will forever be beyond any methodical scientific decision procedures, such as "Does God exist?" or "Which of the two slits in a double-slit interference experiment does a quantum particle pass through?"

Those last two questions are of different natures, although both are unanswerable. Not only is the question as to whether God exists beyond any solid scientific argumentation, but it must be like that. Any other possibility would be the end of religion. If God were provably existent, then the notion of belief would be empty; any religious behavior would be mere opportunism. But what about the quantum question? Which of the two paths does a particle take in a double-slit experiment?

We learned from quantum physics that to answer this kind of question we need to do an experiment allowing us to determine whether the particle takes slit A or slit B. But doing that, we also learned, significantly modifies the experiment itself. Answering the question implies introducing the specific apparatus that allows us to answer it. Introducing an apparatus that permits us to determine which slit a particle takes automatically means that the phenomenon of quantum interference disappears, because of the unavoidable interaction with that apparatus. Or, in the case of the famous Schrödinger cat, asking whether the cat is alive or dead immediately destroys the quantum superposition of the alive and dead states.

Therefore, we have here a completely new situation, not encountered before in science—and probably not in philosophy, either: Creating a situation in which a question can be answered modifies the situation. An experimental quantum setup, or any quantum situation, can represent only a finite amount of information—here, either interference or path information. And it is up to the experimentalist to decide which information is actually existing, real, manifest, in a concrete situation. The experimentalist does this by choosing appropriate apparatus. So information has a fundamental nature of a new kind—a kind not present in classical, nonquantum science.

What does this have to do with the Internet? Today we are busy developing quantum communication over large distances. Using quantum communication links, we will connect future quantum computers that work on a completely new level of complexity, compared with existing computers. To the best of my knowledge, this will be the first time that humanity has developed a technology that has no parallel at all in the known universe (assuming that the functioning of the brain can, in the end, be explained by nonquantum processes).

What will this mean for our communication? It's impossible to tell. It's a prospect even murkier than the predictions about the applications of the laser or the microchip, just to name two more recent examples. We will be entering a completely new world, where information is even more fundamental than it is today. And one can hope that the present irritation experienced by many because of the Internet will turn out to have been just an episode in the development of humanity. But maybe I am being too optimistic.

Edge, A to Z (*Pars Pro Toto*)

Hans Ulrich Obrist

Curator, Serpentine Gallery, London; editor, A Brief History of Curating; Formulas for Now

A is for And

The Internet made me think more BOTH/AND instead of EITHER/OR or NEITHER/NOR.

B is for Beginnings

In terms of my curatorial thinking, my eureka moments occurred pre-Internet, when I met visionary Swiss artists Fischli/Weiss (Peter Fischli and David Weiss) in 1985. These conversations freed me up—freed my thoughts as to what curating could be and how curating can produce reality. The arrival of the Internet was a trigger for me to think more in the form of Oulipian lists—practical-poetical, evolutive, and often nonlinear lists. This A to Z, as you'll see, is an incomplete list . . . Umberto Eco calls the World Wide Web the "mother of all lists," infinite by definition and in constant evolution.

C is for Curating the World

The Internet made me think toward a more expanded notion of curating. Stemming from the Latin *curare*, the word *curating* originally meant "to take care of objects in museums." Curation has long since evolved. Just as art is no longer limited to traditional genres, curating no longer is confined to the gallery or museum but has expanded across all boundaries. The rather obscure and very specialized notion of curating has become much more publicly used—one talks about the curating of Websites—and this marks a very good moment to rediscover the pioneering history of art curating as a toolbox for twenty-first-century society at large.

D is for Delinking

In the years before being online, there were many interruptions by phone and fax day and night. The reality of being permanently linked triggered my increasing awareness of the importance of moments of concentration—moments without interruption that require me to be completely unreachable. I no longer answer the phone at home, and I answer my mobile phone only in the case of fixed telephone appointments. To link is beautiful. To delink is sublime (Paul Chan).

D is also for Disrupted narrative continuity

Forms of film montage, as the disruption of narrative and the disruption of spatial and temporal continuity, have been a staple tactic of the avant-garde from Cubism and Eisenstein through Brecht to Kluge or Godard. For avant-gardism as a whole, it was essential that these tactics be recognized (experienced) as a disruption. The Internet has made disruption and montage the operative bases of everyday experience. Today, these forms of disruption can be harnessed and poeticized. They can foster new connections, new relationships, new productions of reality: reality as life-montage/life as reality-disruption? Not one story but many stories . . .

D is for Doubt

A certain unreliability of technical and material information on the Internet brings us to the notion of doubt. I feel that doubt has become more pervasive. The artist Carsten Höller has invented the Laboratory of Doubt, which is opposed to mere representation. As he has told me, "Doubt and perplexity . . . are unsightly states of mind we'd rather keep under lock and key because we associate them with uneasiness, with a failure of values." Höller's credo is not to do, not to intervene. To exist is to do, and not to do is a way of doing. "Doubt is alive; it paralyzes certainty."

E is for Evolutive exhibitions

The Internet makes me think more about nonfinal exhibitions, exhibitions in a state of becoming. When conceiving exhibitions, I sometimes like to think of randomized algorithms, access, transmission, mutation, infiltration, circulation (the list goes on). The Internet makes me think of exhibitions less as top-down master plans than as bottom-up processes of self-organization.

F is for Forgetting

The ever growing, ever pervasive records that the Internet produces make me think sometimes about the virtues of forgetting. Is a limited life-space of certain information and data becoming more urgent?

H is for Handwriting (and drawing, ever drawing)

The Internet has made me aware of the importance of handwriting and drawing. I typed all my early texts, but the more the Internet has become all-encompassing, the more I have felt that something went missing. Hence the idea to reintroduce handwriting. More and more of my correspondence consists of handwritten letters scanned and sent by e-mail. On a professional note, I observe, as a curator, the importance of drawing in current art production. One can also see it in art schools: a moment when drawing is an incredibly fertile zone.

I is for Identity

"Identity is shifty, identity is a choice" (Etel Adnan).

I is also for Inactual considerations

The future is always built out of fragments of the past. The Internet has brought thinking more into the present tense, raising questions of what it means to be contemporary. Recently, Giorgio Agamben revisited Nietzsche's "Inactual Considerations," arguing that the one who belongs to his or her own time is the one who does not coincide perfectly with it. It is because of this shift, this anachronism, that he or she is more apt than others to perceive and to capture his or her time. Agamben follows this observation with his second definition of contemporaneity: The contemporary is the one who is able to perceive obscurity, who is not blinded by the lights of his or her time or century.

This leads us, interestingly, to the importance of astrophysics in explaining the relevance of obscurity for contemporaneity. The seeming obscurity in the sky is the light that travels to us at full speed but which can't reach us because the galaxies from which it originates are ceaselessly moving away from us at a speed superior to that of light. The Internet and a certain resistance to its present tense have made me increasingly aware that there is an urgent call to be contemporary. To be contemporary means to perpetually come back to a present where we have never yet been. To be contemporary means to resist the homogenization of time, through ruptures and discontinuities.

M is for Maps

The Internet increased the presence of maps in my thinking. It's become easier to make maps, to change them, and also to work on them collaboratively and collectively and share them (e.g., Google Maps and Google Earth). After the focus on social networks of the last couple of years, I have come to see the focus on location as a key dimension.

N is for New geographies

The Internet has fueled (and been fueled by) a relentless economic and cultural globalization, with all its positive and negative aspects. On the one hand, there is the danger of homogenizing forces, which is also at stake in the world of the arts. On the other hand, there are unprecedented possibilities for difference-enhancing global dialogs. In the long duration, there have been seismic shifts, like that in the sixteenth century, when the paradigm shifted from the Mediterranean to the Atlantic. We are living through a period in which the center of gravity is transferring to new centers. The early twenty-first century is seeing the growth of a polyphony of art centers in the East and West as well as in the North and South.

N is also for Nonmediated experiences; N is for the New live

I feel a growing desire for nonmediated experiences. Depending on one's point of view, the virtual may be a new and liberating prosthesis of the body or it may threaten the body. Many visual artists today negotiate and mediate between these two staging encounters of nonmediated intersubjectivity. In the music fields, the crisis of the record industry goes hand in hand with the greater importance of live concerts.

P is for Parallel realities

The Internet creates and fosters new constituencies, new microcommunities. As a system that infinitely breeds new realities, it is predisposed to reproduce itself in a proliferating series of ever more functionally differentiated subsystems. As such, it makes my thinking go toward the production of parallel realities, bearing witness to the multiverse (as the physicist David Deutsch might say), and for better or worse the Internet allows that which is already latent in the "fabric of reality" to unravel itself and expand in all directions.

P is also for Protest against forgetting

I feel an urgency to conduct more and more interviews, to make an effort to preserve traces of intelligence from the last few decades—particularly the testimonies of the twentieth-century pioneers who are in their eighties or nineties (or older) and whom I regularly interview: testimonies about a century from those who are not online and who very often fall into oblivion. This protest might, as Rem Koolhaas has told me, act as "a hedge against the systematic forgetting that hides at the core of the information age and which may in fact be its secret agenda."

S is for Salon of the twenty-first century

The Internet has made me think more about whom I would like to introduce to whom, and about whether to cyberintroduce people or introduce them in person through actual salons for the twenty-first century (see the Brutally Early Club).

The Degradation of Predictability—and Knowledge

Nassim N. Taleb

Distinguished Professor of Risk Engineering, New York University–Polytechnic Institute; principal, Universa Investments; author, The Black Swan

I used to think the problem of information is that it turns *Homo sapiens* into fools—we gain disproportionately in confidence, particularly in domains where information is wrapped in a high degree of noise (say, epidemiology, genetics, economics, etc.). So we end up thinking we know more than we do, which, in economic life, causes foolish risk taking. When I started trading, I went on a news diet and I saw things with more clarity. I also saw how people built too many theories based on sterile news, fooled by the randomness effect. But things are a lot worse. Now I think that, in addition, the supply and spread of information turns the world into Extremistan (a world I describe as one in which random variables are dominated by extremes, with Black Swans playing a large role in them). The Internet, by spreading information, causes an increase in interdependence, the exacerbation of fads (bestsellers like *Harry Potter* and runs on banks become planetary). Such a world is more "complex," more moody, much less predictable.

So consider the explosive situation: More information (particularly thanks to the Internet) causes more confidence and illusions of knowledge while degrading predictability.

Look at the economic crisis that started in 2008, There are about a million persons on the planet who identify themselves as in the field of economics. Yet just a handful realized the possibility and depth of what could take place and protected themselves from the consequences. At no time in the history of humankind have we lived in so much ignorance (easily measured in terms of forecast errors) coupled with so much intellectual hubris. At no point have we had central bankers missing elementary risk metrics—like debt levels, which even the Babylonians understood well.

I recently talked to a scholar of rare wisdom and erudition, Jon Elster, who, upon exploring themes from social science, integrates insights from all authors in the corpus of the past twenty-five hundred years, from Cicero and Seneca to Montaigne and Proust. He showed me how Seneca had a very sophisticated understanding of loss aversion. I felt guilty for the time I spent on the Internet. Upon getting home, I found in my mail a volume of posthumous essays by Bishop Pierre Daniel Huet, called *Huetiana*, put together by his admirers circa 1722. It is saddening to realize that, having been born nearly four centuries after Huet, and having done most of my reading with material written after his death, I am not much more advanced in wisdom than he was. Moderns at the upper end are no wiser than their equivalent among the ancients; if anything, they are much less refined.

So I am now on an Internet diet, in order to understand the world a bit better—and make another bet on horrendous mistakes by economic policy makers. I am not entirely deprived of the Internet; this is just a severe diet with strict rationing. True, technologies are the greatest things in the world, but they have far too monstrous side effects—and ones rarely seen ahead of time. And since I have been spending time in the silence of my library with little informational pollution, I can feel harmony with my genes; I feel I am growing again.

Calling You on Your Crap

Sean Carroll

Theoretical physicist, Caltech; author, From Eternity to Here: The Quest for the Ultimate Theory of Time

I wanted to write that the Internet keeps people honest. The image of thousands of readers bursting into laughter gave me pause.

So let me put it this way: The Internet helps enable honesty. Many of us basically want to be honest, but we're fighting all sorts of other impulses—the desire to appear clever or knowledgeable, to support a point we're trying to make, to feel the satisfaction of a rant well ranted. In everyday conversation, when we know something specific about the expertise and inclinations of our audience, these impulses may tempt us into laziness: pushing a point too hard, claiming as fact some anecdote whose veracity isn't completely reliable. We're only human.

Nothing highlights our natural tendencies to exaggerate and overclaim quite like a widely distributed, highly interconnected communication network with nearly instantaneous feedback. There is no shortage of overblown and untrue claims on the Internet. But for those of us who would really like to be as honest and accurate as is reasonably possible, the Internet is an invaluable corrective.

All else being equal, it is a virtue to know true things. But there is also the virtue of assigning accurate degrees of confidence to the things we think we know. There are some things I have studied personally and in depth, such that I have acquired some expertise; there are other things I've read somewhere, or heard from a friend, that sound pretty reasonable. And there are still other things that wouldn't sound at all reasonable to an objective observer but that line up with my cherished beliefs. Distinguishing between these categories is a major part of being intellectually honest.

Engaging with ideas online—stating what I believe, arguing in favor of it to the best of my ability, and stretching my mind by reading things outside my comfort zone—is immensely helpful in separating well-established facts from wishful thinking. The thing about the Internet is, people will call you on your crap. Even if I don't know exactly what I'm talking about, somebody out there does. On discussion boards, in blog comment threads, on Websites of colleagues or students on another continent, if I say something that manages to be interesting but wrong, chances are someone will set me straight. Not that everyone necessarily listens. It's my responsibility to be open enough to listen to the critiques and improve my position—but that's always been my job. The Internet merely helps us along.

The distinction is not only between the Internet and sitting around a table having a bull session with your friends; it applies to conventional print media as well, from books to newspapers and magazines. Sure, someone can write a book review or pen a strident letter to the editor. But time scales matter. If I put up a blog post in the morning and get several comments before lunchtime along the lines of "That's about as wrong as anything I've ever seen you write" or "What were you thinking?" complete with links to sources that set me straight, it's difficult to simply pretend I don't notice.

I once heard, as an example of how online communication was degrading our discourse by drowning us in lies and misinformation, the crazy claim that Stephen Hawking wouldn't have been cared for under the United Kingdom's National Health Service—which, of course, is exactly what did care for him, thus offering an unusually juicy self-refutation. But bringing up this example as a criticism of the Internet is equally self-refuting. The initial lie didn't appear online but in a good, old-fashioned newspaper. Twenty years ago, that's as far as it would have circulated after making a brief impression in the minds of its readers. But today countless online sources leapt to make fun of the ridiculous lengths to which opponents of health care reform were willing to go. Perhaps next time the editorial writers will be more careful in their choice of colorful counterfactuals.

All of which is incredibly small potatoes, of course. The Internet in its current configuration is only a hint of what we'll have a hundred years from now; feel free to visualize your own favorite chip-in-your-head scenario. Cutting down on the noise will ultimately be just as great a challenge as connecting to the signal. But even now the Internet is a great help to those of us who prefer to be kept honest—it's just up to us to take advantage.

How I Think About How I Think

Lera Boroditsky

Assistant professor of psychology, Stanford University

Consider a much earlier piece of technology than the Internet: the fork. When I take a fork (or any tool) in my hand, the multimodal neurons in my brain tracking my hand's position immediately expand their receptive fields. They start to keep track of a larger part of space, expanding their view to include perhaps that succulent morsel of lamb that is now within my fork's reach. My brain absorbs the tool in my hand into the very representation of my physical self; the fork is now, in an important neural sense, a part of my body. (In case absorbing a fork into your sense of self seems strange, it may help to note that this phenomenon was discovered by a former dentist who ingeniously trained rhesus monkeys to search for food with tools suspiciously resembling dental endoscopes.) If grabbing a humble fork can expand my neurons' receptive fields, imagine what happens when I grab a mouse and open a Web browser.

Indeed, research in the last decade has shown that our brains change, grow, and adapt dramatically as we engage with the world in new ways. London taxi drivers grow larger hippocampi (a part of the brain heavily involved in navigation) as they gain knowledge maneuvering through the maze of London streets. Playing video games significantly improves people's spatial attention and object-tracking abilities, giving a regular schmo the attentional skills of a fighter pilot. At this rate, we'll be lucky if the list of basic drives controlled by the hypothalamus—the famous four F's: *fighting*, *fleeing*, *feeding*, and *how's your father?*—doesn't soon need to be augmented with a fifth, *Facebook*. This, by the way, is the reason I give for not joining social networking sites. My hypothalamus has more important business to attend to, thanks!

My favorite human technologies are the ones we no longer even notice as technologies—they just seem like natural extensions of our minds. Numbers are one such example, a human-invented tool that, once learned, has incredible productive power in the mind. Writing is another. It no longer seems magical, in the literate world, to communicate a complex set of thoughts silently across vast reaches of time and space using only a cocktail napkin and some strategically applied stains. Yet being able to write things down, draw diagrams, and otherwise externalize the contents of our minds into some stable format has drastically augmented our cognitive and communicative abilities. By far the most amazing technological marvels that humans ever created (and what I spend most of my time thinking about) are the languages we speak. Now, there's an immensely complex tool that really changed things for us humans. You think keeping up a correspondence with friends was hard before e-mail? Well, you should have tried it before language! Importantly, the particulars of the languages we speak have shaped not only how we communicate our thoughts but also the very nature of the thoughts themselves.

There are, of course, facile or insipid ways of construing the nature of human thought such that "the way you think" isn't, and can't, be changed by technology. For example, I could define the basic mechanisms of thought as "neurons fire, at different times, some more than others, and that is how I think." Well, all right, that's technically true, and the Internet is not changing that. But on any more interesting or useful construal of human thought, technology has been shaping us for as long as we've been making it.

More than shaping how I think, the Internet is also shaping how I think about how I think. Scholars interested in the nature of mind have long relied on technology as a source of metaphors for explaining how the mind works. First the mind was a clay tablet, then an abacus, a calculator, a telephone switchboard, a computer, a network. These days, new tools continue to provide convenient (perhaps in the

7-Eleven sense of convenient, as in nearby but ultimately unsatisfying) metaphors for explaining the mind. Consciousness, for example, is not unlike Twitter—millions of mundane messages bouncing around, all shouting over one another, with only a few rising as trending topics. Take that, Dan Dennett! Consciousness explained, in 140 characters or less!

I Am Not Exactly a Thinking Person—I Am a Poet

Jonas Mekas

Filmmaker, critic; cofounder, Film-Makers' Cooperative, Film-Makers' Cinematheque, Anthology Film Archives

I am a farmer boy. When I grew up, there was only one radio in our entire village of twenty families. And, of course, no TV, no telephone, and no electricity. I saw my first movie when I was fourteen. In New York, in 1949, I fell in love with cinema. In 1989, I switched to video. In 2003, I embraced computer/Internet technologies.

I am telling you this to indicate that my thinking is now only entering the Internet Nation. I am not really thinking yet the Internet way—I am only babbling.

But I can tell you that the Internet has already affected the content, form, and working procedures of everything I do. It's entering my mind secretly, indirectly.

In 2007, I did a 365 Day Project. I put one short film on the Internet every day. In cinema, when I was making my films, the process was abstract. I could not think about the audience; I knew the film would be placed in a film distribution center and eventually someone would look at it. In my 365 Day Project, I knew that later, the same day, I would put it on the Internet and within minutes it would be seen by all my friends—and strangers, too—all over the world. So I felt as if I were conversing with them. It's intimate. It's poetic. I am not thinking anymore about problems of distribution; I am just exchanging my work with some friends. Like being part of a family. I like that. It makes for a different state of mind. Whether a state of mind has anything or nothing to do with thinking, that's unimportant to me. I am not exactly a thinking person. I am a poet.

I would like to add one more note about what the Internet has done to me. And that is that I began paying more attention to everything the Internet seems to be eliminating. Books, especially. But also nature. In short, the more it all expands into virtual reality, the more I feel a need to love and protect actual reality. Not because of sentimental reasons, no—from a very real, practical, almost a survival need, from my knowledge that I would lose an essential part of myself by losing actual reality, both cultural and physical.

Kayaks Versus Canoes

George Dyson

Science historian; author, Darwin Among the Machines

In the North Pacific, there were two approaches to boatbuilding. The Aleuts and their kayak-building relatives lived on barren, treeless islands and built their vessels by piecing together skeletal frameworks from fragments of beachcombed wood. The Tlingit and their dugout-canoe-building relatives built their vessels by selecting entire trees out of the rain forest and removing wood until there was nothing left but a canoe.

The Aleut and the Tlingit achieved similar results—maximum boat, minimum material—by opposite means. The flood of information unleashed by the Internet has produced a similar cultural split. We used to be kayak builders, collecting all available fragments of information to assemble the framework that kept us afloat. Now we have to learn to become dugout-canoe builders, discarding unnecessary information to reveal the shape of knowledge hidden within.

I was a hardened kayak builder, trained to collect every available stick. I resent having to learn the new skills. But those who don't will be left paddling logs, not canoes.

The Upload Has Begun

Sam Harris

Neuroscientist; chairman, Project Reason; author, The Moral Landscape: How Science Can Determine Human Values

It is now a staple of scientific fantasy, or nightmare, to envision that human minds will one day be uploaded onto a vast computer network like the Internet. While I am agnostic about whether we will ever break the neural code, allowing our inner lives to be read out as a series of bits, I notice that the prophesied upload is slowly occurring in my own case. For instance, the other day I recalled a famous passage from Adam Smith that I wanted to cite: something about an earthquake in China. I briefly considered scouring my shelves in search of my copy of *The Wealth of Nations.* But I have thousands of books spread throughout my house, and they are badly organized. I recently spent an hour looking for a title, and then another skimming its text, only to discover that it wasn't the book I had wanted in the first place. And so it would have proved in the present case, for the passage I dimly remembered from Smith is to be found in *The Theory of Moral Sentiments.* Why not just type the words "adam smith china earthquake" into Google? Mission accomplished.

Of course, more or less everyone has come to depend on the Internet in this way. Increasingly, however, I rely on Google to recall my own thoughts. Being lazy, I am prone to cannibalizing my work: Something said in a lecture will get plowed into an op-ed; the op-ed will later be absorbed into a book; snippets from the book may get spoken in another lecture. This process will occasionally leave me wondering just how and where and to what shameful extent I have plagiarized myself. Once again, the gates of memory swing not from my own medial temporal lobes but from a computer cluster far away, presumably where the rent is lower.

This migration to the Internet now includes my emotional life. For instance, I occasionally engage in public debates and panel discussions where I am pitted against some over-, under-, or miseducated antagonist. "How did it go?" will be the question posed by wife or mother at the end of the day. I now know that I cannot answer this question unless I watch the debate online, for my memory of what happened is often at odds with the later impression I form based upon seeing the exchange. Which view is closer to reality? I have learned to trust the YouTube version. In any case, it is the only one that will endure.

Increasingly, I develop relationships with other scientists and writers that exist entirely online. Jerry Coyne and I just met for the first time, in a taxi in Mexico. But this was after having traded hundreds of e-mails. Almost every sentence we have ever exchanged exists in my Sent folder. Our entire relationship is therefore searchable. I have many other friends and mentors who exist for me in this way, primarily as e-mail correspondents. This has changed my sense of community profoundly. There are people I have never met who have a better understanding of what I will be thinking tomorrow than some of my closest friends do.

And there are surprises to be had in reviewing this digital correspondence. I recently did a search of my Sent folder for the phrase "Barack Obama" and discovered that someone had written to me in 2004 to say that he intended to give a copy of my first book to his dear friend Barack Obama. Why didn't I remember this exchange? Because at the time I had no idea who Barack Obama was. Searching my bitstream, I am reminded not only of what I used to know but also of what I never properly understood.

I am by no means infatuated with computers. I do not belong to any social networking sites; I do not tweet (yet); and I do not post images to Flickr. But even in my case, an honest response to the Delphic admonition "Know thyself" already requires an Internet search.

Hell if I Know

Gregory Paul

Independent researcher; author, Dinosaurs of the Air: The Evolution and Loss of Flight in Dinosaurs and Birds

Being among those who have predicted that humans will be uploading their minds into cybermachines in the not too distant future, I'm enthusiastic about the Internet. But the thinking of my still-primate mind about the new mode of information exchange is more ambiguous.

No doubt the Internet is changing the way I operate and influence the world around me. Type "gregory paul religion society" into Google and nearly 3.5 million hits come up. I'm not entirely sure what that means, but it looks impressive. An article in a British newspaper on my sociological research garnered more than seven hundred comments. Back in the twentieth century, I could not imagine my technical research making such an impression on the global sociopolitical scene, because the responsible mechanism—publishing in open-access online academic journals—was not available. The new communication environment is undoubtedly altering my research and publicity strategy relative to what it would be in a less digital world. Even so, I am not entirely sure how my own actions are being modified. The only way to find out would be to run a parallel-universe experiment, in which everything is the same except for the existence of Internet-type communications, and see what I do.

What is disturbing to this human raised on hard-copy information transmission is how fast the Internet is destroying a large part of it. My city no longer has a truly major newspaper, and the edgy, free *City Paper* is a pale shadow of its former self. I've enjoyed living a few blocks from a major university library, because I could browse through the extensive journal stacks, leafing through assorted periodicals to see what was up in the latest issues. Because the search was semi-random, it was often pleasantly and usefully serendipitous. Now that the Hopkins library has severely cut back on paper journals as the switch to online continues, it is less fun. Looking up a particular article is often easier online (and it's nice to save trees), but checking the contents of the latest issue of *Geology* on the library computer is neither as pleasant nor as convenient. I suspect the range of my information intake has narrowed, and that can't be good.

On the positive side, it could be amazingly hard to get basic information before the Web showed up. In my teens, I was fascinated by the destruction of the HMS *Hood* in 1941 but unable to get a clear impression of the famed vessel's appearance for a couple of years, until I saw a friend's model, and I did not see a clear image until well after that. Such extreme data deprivation is over, thanks to Wikipedia et cetera. But even the Internet cannot fill all information gaps. It's often difficult to search out obscure details of the sort found only in books, which can look at subjects in depth. Websites reference books, but if the Internet limits the production of manuscript-length works, then the quality of information is going to suffer.

As for the question of how the Internet is changing my thinking, online apps facilitate the statistical analyses that are expanding my sociological interests and conclusions, leading to unanticipated answers to some fundamental questions about popular religion that I am delighted to uncover. Beyond that, there are more subtle effects, but exactly what they are I am not sure, sans the parallel-world experiment. I fear that the brevity favored by screen versus page is shortening my attention span. It's as if one of Richard Dawkins's memes had altered my unwilling mind like a bad science fiction story. But that's a nonquantitative, anecdotal impression; perhaps I just *think* my thinking has changed. It may be that the new arrangement is not altering my mental exertions further than it is because the old-fashioned mind generated by my brain remains geared to the former system.

The new generation growing up immersed in the digital complex may be developing thinking processes more suitable for the new paradigm, for better or worse. But perhaps human thinking is not as amenable to modification by external factors as one might expect. And the Internet may be more retro than it seems. The mass media of the twentieth century were truly novel, because the analog-based technology turned folks from home entertainers and creators (gathering around the piano and singing and composing songs and the like) to passive consumers of a few major outlets (sitting around the telly and fighting over the remote). People are using hyperfast digital technology to return to self-creativity and entertainment. How all this is affecting young psyches is a matter for sociobehavioral and neuropsychological research to sort out.

But how humans old and young are affected may not matter all that much. In the immediacy of this early-twenty-first-century moment, the Internet revolution may look more radical than it actually is. It could merely be an introduction to the real revolution. The human domination of digital communications will be a historically transitory event if and when high-level-thinking cyberminds start utilizing the system. The ability of superintelligences to share and mull over information will dwarf what mere humans can manage. Exactly how will the interconnected *über*minds think? Hell if I know.

What I Notice

Brian Eno

Artist; composer; recording producer: U2, Coldplay, Talking Heads, Paul Simon; recording artist

I notice that some radical social experiments that would have seemed utopian to even the most idealistic anarchist fifty years ago are now working smoothly and without much fuss. Among these are open-source development, shareware and freeware, Wikipedia, MoveOn, and UK Citizens Online Democracy.

I notice that the Net didn't free the world in quite the way we expected. Repressive regimes can shut it down, and liberal ones can use it as a propaganda tool. On the upside, I notice that the variable trustworthiness of the Net has made people more skeptical about the information they get from all other media.

I notice that I now digest my knowledge as a patchwork drawn from a wider range of sources than I used to. I notice, too, that I am less inclined to look for joined-up, finished narratives and more inclined to make my own collage from what I can find. I notice that I read books more cursorily, scanning them in the same way that I scan the Net—"bookmarking" them.

I notice that the turn-of-the-century dream of bioethicist Darryl Macer to make a map of all the world's concepts is coming true autonomously—in the form of the Net.

I notice that I correspond with more people but at less depth. I notice that it is possible to have intimate relationships that exist only on the Net and have little or no physical component. I notice that it is even possible to engage in complex social projects, such as making music, without ever meeting your collaborators. I am unconvinced of the value of these.

I notice that the idea of "community" has changed: Whereas that term once connoted some sort of physical and geographical connectedness between people, now it can mean "the exercise of any shared interest." I notice that I now belong to hundreds of communities—the community of people interested in active democracy, the community of people interested in synthesizers, in climate change, in Tommy Cooper jokes, in copyright law, in a cappella singing, in loudspeakers, in pragmatist philosophy, in evolution theory, and so on.

I notice that the desire for community is sufficiently strong for millions of people to belong to entirely fictional communities, such as Second Life and World of Warcraft. I worry that this may be at the expense of First Life.

I notice that more of my time is spent in words and language—because that is the currency of the Net—than it was before. My notebooks take longer to fill. I notice that I mourn the passing of the fax machine, a more personal communication tool than e-mail because it allowed the use of drawing and handwriting. I notice that my mind has reset to being primarily linguistic rather than, for example, visual.

I notice that the idea of "expert" has changed. An expert used to be "somebody with access to special information." Now, since so much information is equally available to everyone, the idea of "expert" becomes "somebody with a better way of interpreting." Judgment has replaced access.

I notice that I have become a slave to connectedness—that I check my e-mail several times a day, that I worry about the heap of unsolicited and unanswered mail in my inbox. I notice that I find it hard to get a whole morning of uninterrupted thinking. I notice that I am expected to answer e-mails immediately, and that it is difficult not to. I notice that as a result I am more impulsive.

I notice that I more often give money in response to appeals made on the Net. I notice that memes can now spread like virulent infections through the vector of the Net—and that this isn't always good.

I notice that I sometimes sign petitions about things I don't really understand, because it's easy. I assume that this kind of irresponsibility is widespread.

I notice that everything the Net displaces reappears somewhere else in a modified form. For example, musicians used to tour to promote their records, but since records stopped making much money due to illegal downloads, they now make records to promote their tours. Bookstores with staff who know about books, and record stores with staff who know about music, are becoming more common.

I notice that as the Net provides free or cheap versions of things, the "authentic experience"—the singular experience enjoyed without mediation—becomes more valuable. I notice that more attention is given by creators to the aspects of their work that can't be duplicated. The "authentic" has replaced the reproducible.

I notice that hardly any of us have thought about the chaos that would ensue if the Net collapsed.

I notice that my daily life has been changed more by my mobile phone than by the Internet.

It's Not What You Know, It's What You Can Find Out

Marissa Mayer

Vice president, Search Products and User Experience, Google

It's not what you know, it's what you can find out. The Internet has put resourcefulness and critical thinking at the forefront and relegated memorization of rote facts to mental exercise or enjoyment. Because of the abundance of information and this new emphasis on resourcefulness, the Internet creates a sense that anything is knowable or findable—as long as you can construct the right search, find the right tool, or connect to the right people. The Internet enables better decision making and a more efficient use of time.

Simultaneously, it also leads to a sense of frustration when the information doesn't exist online. (What do you *mean*, the store hours aren't anywhere? Why can't I see a particular page of this book? And if not verbatim, has no one quoted this even in *part*? What do you *mean* that page isn't available?)

The Internet can facilitate an incredible persistence and availability of information, but given the Internet's adolescence, not all the information is there yet. I find that in some ways my mind has adopted this new way of the thinking—that is, relying on information's existence and availability—so completely that it's almost impossible to conclude that the information isn't findable because it simply isn't online.

The Web has also enabled amazing dynamic visualizations, whereby an ideal presentation of information is constructed—a table of comparisons, or a data-enhanced map, for example. These visualizations—news from around the world displayed on a globe, say, or a sortable table of airfares—can greatly enhance our comprehension of the world and our sense of opportunity. We can grasp in an instant what would have taken months to create just a few short years ago. Yet the Internet's lack of structure means that these types of visualization can't be constructed for all data. To achieve true automated general understanding and visualization, we will need much better machine learning, entity extraction, and semantics capable of operating at vast scales.

On that note—and regarding future Internet innovation—the important question may be not how the Internet is changing the way we think but how the Internet is teaching itself to think.

When I'm on the Net, I Start to Think

Ai Weiwei

Artist; curator; architectural designer, Beijing National Stadium (the Bird's Nest); cultural and social commentator; activist

Nowadays I mostly think only on the Internet. My thinking is divided into on the Net and off the Net. If I'm not on the Net, I don't think that much; when I'm on the Net, I start to think. In this way, my thinking becomes part of something else.

The Internet Has Become Boring

Andrian Kreye

Editor, The Feuilleton *(arts and essays) of the German daily* Süddeutsche Zeitung, *Munich*

I think faster now. The Internet has somewhat freed me of some of the twentieth century's burdens: the burden of commuting, the burden of coordinating communication, the burden of traditional literacy. I don't think the Internet would be of much use if I hadn't carried those burdens to excess all through my life. If continually speeding up thinking constitutes changing the way I think, though, the Internet has done a marvelous job.

I wasn't an early adopter, but the process of adaptation started early. I didn't yet understand what would come upon us when, one afternoon in 1989 at MIT's Media Lab, Marvin Minsky told me that the most important trait of a computer would be not its power but what it would be connected to. A couple of years later, I stumbled upon the cyberpunk scene in San Francisco. People were popping "smart" drugs (which didn't do anything), Timothy Leary declared virtual reality the next psychedelic (which never panned out), Todd Rundgren warned of a coming overabundance of creative work without a parallel rise in great ideas (which is now reflected in laments about the rise of the amateur). It was still the old underground, running the new emerging culture. This new culture was driven by thought rather than art, though. That's also where I met Cliff Figallo, who ran a virtual community called the WELL. He introduced me to John Perry Barlow, who had just established the Electronic Frontier Foundation. The name said it all: There was a new frontier.

It would still take me a few more years to grasp this. One stifling evening in a rented apartment in downtown Dakar, my photographer and I disassembled a phone line and a modem to circumvent incompatible jacks and get our laptop to dial up some node in Paris. It probably saved us a good week of research in the field. Now my thinking started to take on the speed I had sensed in Boston and San Francisco. Continually freeing me of the aforementioned burdens, it has allowed me to focus even more on the tasks expected of me as a journalist—finding context, meaning, and a way to communicate complex topics as simply as possible.

An important development—one that has allowed this new freedom—is that possibly the biggest thing that's happened to the Internet over the past few years is that it's become boring. Gone are the adventurous days of using a pocketknife to log on to Paris from Africa. Even in remote places on this planet, logging on to the Net means merely turning on your machine. This paradigm reigns all through the Web. Twitter is one of the simplest Internet applications ever developed. Still, it has sped up my thinking in ever more ways. Facebook in itself is dull, but it has created new networks not possible before. Integrating all media into a blog has become so easy that grammar school kids can do it, so that the free-form forum has become a great place to test out new possibilities. I don't think about the Internet anymore. I just use it.

All this, however, might not constitute a change in thinking. I haven't changed my mind or my convictions because of the Internet. I haven't had any epiphanies while sitting in front of a screen. The Internet, so far, has given me no memorable experiences, although it may have helped usher some along. People, places, and experiences are what change the way I think.

The Dumb Butler

Joshua Greene

Cognitive neuroscientist and philosopher, Harvard University

Have you ever read a great book published before the mid-1990s and thought, "My goodness! These ideas are so primitive! So . . . pre-Internet!"? Me neither. The Internet hasn't changed the way we think any more than the microwave oven has changed the way we digest food. The Internet has provided us with unprecedented access to information, but it hasn't changed what we do with it once it's made it into our heads. This is because the Internet doesn't (yet) know how to think. We still have to do it for ourselves, and we do it the old-fashioned way.

One of the Internet's early disappointments was the now defunct Website Ask Jeeves. (It was succeeded by Ask.com, which dropped Jeeves in 2006.) Jeeves appeared as a highly competent infobutler, who could understand and answer questions posed in natural language ("How was the East Asian economy affected by the Latin American debt crisis?" "Why do fools fall in love?"). Anyone who spent more than a few minutes querying Jeeves quickly learned that Jeeves himself didn't understand squat. Jeeves was just a search engine like the rest, mindlessly matching the words contained in your question to words found on the Internet. The best Jeeves could do with your profound question—the best any search engine can do today—is direct you to the thoughts of another human being who has already attempted to answer a question related to yours. This is not to say that cultural artifacts can't change the way we think.

The political philosopher Jim Flynn has documented substantial gains in IQ in the twentieth century (the Flynn effect), which he attributes to our enhanced capacity for abstract thought, which he in turn attributes to the cognitive demands of the modern marketplace. Why hasn't the Internet had a comparable effect? The answer, I think, is that the roles of master and servant are reversed. We place demands on the Internet, but the Internet hasn't placed any fundamentally new demands on us. In this sense, the Internet really *is* like a butler. It gives us the things we want faster and with less effort, but it doesn't give us anything we couldn't otherwise get for ourselves, and it doesn't require us to do anything more than give comprehensible orders.

Someday we'll have a nuts-and-bolts understanding of complex abstract thought, which will enable us to build machines that can do it for us and perhaps do it better than we do—and perhaps teach us a thing or two about it. But until then, the Internet will continue to be nothing more, and nothing less, than a very useful, and very dumb, butler.

Finding Stuff Remains a Challenge

Philip Campbell

Editor-in-chief, Nature

For better or worse, the Internet is changing *when* I think—nighttime ideas can be instantly acted on. But much more important, the Internet has immeasurably supported my breadth of consideration and enhanced my speed of access to relevant stuff. Frustrations arise, above all, where these are constrained—and there's a rub.

We are in sight of technologies that can truly supersede paper, retaining the portability, convenience, and format variety of that medium. Instant payment for added-value content will become easier and, indeed, will be taken for granted in many contexts.

But finding the stuff will remain a challenge. Brands, both publishers' and others', if deployed in a user-friendly way, will by their nature assist those seeking particular types of content. But content within established brands is far from an adequate representation of what matters, and that's why robust and inclusive indexing systems are so important.

I remain uneasy that biologists worldwide are so dependent on a literature-indexing system wholly funded by U.S. taxpayers: PubMed. Nevertheless, it's extraordinarily valuable and works in the interests of researchers—and also of publishers, by making their work accessible without undermining their business models.

I emphasize that last point with good reason. One of the worst (i.e., self-defeatingly shortsighted) acts of "my" industry occurred in the early 2000s. Congress, lobbied by publishers and seemingly ignorant of the proven virtues of PubMed, rejected support for an equivalent search infrastructure, PubSCIENCE, established by the U.S. Department of Energy as an index for physical sciences and energy research. The lobbyists argued, wrongheadedly, that it competed with private-sector databases. It was abandoned in 2002. Publishers have lost opportunities as a result, as has everyone else. Energy research, after all, has never been more urgent nor more in the United States' and the world's interest.

PubMed imposes overly conservative restrictions on what it will index, but it is a beacon nevertheless. Anyone in the natural sciences who, like me, has taken an active interest in the social sciences knows how hopelessly unfindable by comparison is that literature, distributed as it is among books, reports, and unindexed journals. Google Scholar is in some ways valuable, providing access also to some "gray" literatures, but its algorithms are a law unto themselves and, in my experience, miss some of the literature. And so often the books and reports are themselves difficult to obtain.

There are foundations and other funders potentially more enlightened than Congress when it comes to supporting literature digitization and indexing. And universities are developing online repositories of their outputs, though with limited success.

Whatever works! Those wishing to promote the visibility and, dare one say, usefulness of their own work and of their disciplines should hotly pursue online availability of all types of substantive texts and, crucially, inclusive indexing.

Attention, Crap Detection, and Network Awareness

Howard Rheingold

Communications expert; author, Smart Mobs

Digital media and networks can empower only the people who learn how to use them—and pose dangers to those who don't know what they are doing. Yes, it's easy to drift into distraction, fall for misinformation, allow attention to fragment rather than focus, but those mental temptations pose dangers only for the untrained mind. Learning the mental discipline to use thinking tools without losing focus is one of the prices I am glad to pay to gain what the Web has to offer.

Those people who do not gain fundamental literacies of attention, crap detection, participation, collaboration, and network awareness are in danger of all the pitfalls critics point out—shallowness, credulity, distraction, alienation, addiction. I worry about the billions of people who are gaining access to the Net without the slightest clue about how to find knowledge and verify it for accuracy, how to advocate and participate rather than passively consume, how to discipline and deploy attention in an always-on milieu, how and why to use those privacy protections that remain available in an increasingly intrusive environment.

The realities of my life as a professional writer—if the words didn't go out, the money didn't come in—drove me to evolve a set of methods and disciplines. I know that others have mastered far beyond my own practice the mental habits I've stumbled upon, and I suspect that learning these skills is less difficult than learning long division. I urge researchers and educators to look more systematically where I'm pointing.

When I started out as a freelance writer in the 1970s, my most important tools were a library card, a typewriter, a notebook, and a telephone. In the early 1980s, I became interested in the people at Xerox Palo Alto Research Center (PARC) who were using computers to edit text without physically cutting, pasting, and retyping pages. Through PARC, I discovered Douglas Engelbart, who had spent the first decade of his career trying to convince somebody, anybody, that using computers to augment human intellect was not a crazy idea. Engelbart set out in the early 1960s to demonstrate that computers could be used to automate low-level cognitive support tasks, such as cutting, pasting, and revising text, and also to enable intellectual tools, such as the hyperlink, that weren't possible with Gutenberg-era technology.

He was convinced that this new way to use computers could lead to

> increasing the capability of a man to approach a complex problem situation, to gain comprehension to suit his particular needs, and to derive solutions to problems. Increased capability in this respect is taken to mean a mixture of the following: more-rapid comprehension, better comprehension, the possibility of gaining a useful degree of comprehension in a situation that previously was too complex, speedier solutions, better solutions, and the possibility of finding solutions to problems that before seemed insolvable.*

Important caveats and unpredicted side effects notwithstanding, Engelbart's forecasts have come to pass in ways that surprised him. What did not surprise him was the importance of both the know-how and how-to-know that unlock the opportunities afforded by augmentation technology.

---

* D. C. Engelbart, "Augmenting Human Intellect: A Conceptual Framework," Summary Report AFOSR-3233, Stanford Research Institute, Menlo Park, Calif., October 1962.

From the beginning, Engelbart emphasized that the hardware and software created at his Stanford Research Institute laboratory, from the mouse to the hyperlink to the word processor, were part of a system that included "humans, language, artifacts, methodology, and training." Long before the Web came along, Engelbart was frustrated that so much progress had been made in the capabilities of the artifacts but so little study had been devoted to advancing the language, methodology, and training—the literacies that necessarily accompany the technical capabilities.

Attention is the fundamental literacy. Every second I spend online, I make decisions about where to spend my attention. Should I devote any mind share at all to this comment or that headline?—a question I need to answer each time an attractive link catches my eye. Simply becoming aware of the fact that life online requires this kind of decision making was my first step in learning to tune a fundamental filter on what I allow into my head—a filter that is under my control only if I practice controlling it. The second level of decision making is whether I want to open a tab on my browser because I've decided this item will be worth my time tomorrow. The third decision: Do I bookmark this site because I'm interested in the subject and might want to reference it at some unspecified future time? Online attention taming begins with what meditators call mindfulness—the simple, self-influencing awareness of how attention wanders.

Life online is not solitary. It's social. When I tag and bookmark a Website, a video, an image, I make my decisions visible to others. I take advantage of similar knowledge curation undertaken by others when I start learning a topic by exploring bookmarks, find an image to communicate an idea by searching for a tag. Knowledge sharing and collective action involve collaborative literacies.

Crap detection—Hemingway's name for what digital librarians call credibility assessment—is another essential literacy. If all schoolchildren could learn one skill before they go online for the first time,

I think it should be the ability to find the answer to any question and the skills necessary to determine whether the answer is accurate or not.

Network awareness, from the strength of weak ties and the nature of small-world networks to the power of publics and the how and why of changing Facebook privacy settings, would be the next literacy I would teach, after crap detection. Networks aren't magic, and knowing the principles by which they operate confers power on the knowledgeable. How could people *not* use the Internet in muddled, frazzled, fractured ways, when hardly anybody instructs anybody else about how to use the Net salubriously? It is inevitable that people will use it in ways that influence how they think and what they think.

It is not inevitable that these influences will be destructive. The health of the online commons will depend on whether more than a tiny minority of Net users become literate Netizens.

Information Metabolism

Esther Dyson

Catalyst, information technology start-ups, EDventure Holdings; former chairman, Electronic Frontier Foundation and ICANN; author, Release 2.1

I love the Internet. It's a great tool precisely because it is so content- and value-free. Anyone can use it for his own purposes, good or bad, big or small, trivial or important. It impartially transmits all kinds of content, one-way or two-way or broadcast, public or private, text or video or sound or data.

But it does have one overwhelming feature: immediacy. (And when the immediacy is ruptured, its users gnash their teeth.) That immediacy is seductive: You can get instant answers, instant responses. If you're lonely, you can go online and find someone to chat with. If you want business, you can send out an e-mail blast and get at least a few responses—a .002 percent response rate means 200 messages back (including some hate mail) for a small list. If you want to do good, there are thousands of good causes competing for your attention at the click of your mouse.

But sometimes I think much of what we get on the Internet is empty calories. It's sugar—short videos, pokes from friends, blog posts, Twitter posts (even blogs seem long-winded now), pop-ups, visualizations . . . Sugar is so much easier to digest, so enticing—and ultimately it leaves us hungrier than before.

Worse than that, over a long period many of us are genetically disposed to lose our ability to digest sugar if we consume too much of it. It makes us sick long-term, as well as giving us indigestion and hypoglycemic fits. Could that be true of information sugar as well? Will we become allergic to it even as we crave it? And what will serve as information insulin?

In the spirit of brevity if not immediacy, I leave it to the reader to ponder these questions.

Ctrl + Click to Follow Link

George Church

Professor, Harvard University; director, Personal Genome Project.

If time did permit, I'd begin with the "How" of "How is the Internet changing the way you think?" Not "how much?" or "in what manner?" but "for what purpose?" "To be, that is the question."

Does the Internet pose an existential risk to all known intelligence in the universe or a path to survival? Yes; we see sea change from I-Ching to e-Change.

Yes; it (IT) consumes 100 billion watts, but this is only 0.7 percent of human power consumption.

Yes; it might fragment the attention span of the Twitter generation. (For my world, congenitally shattered by narcolepsy and dyslexia, reading/chatting online in 1968 was no big deal.)

Before cuneiform, we revered the epic poet. Before Gutenberg, we exalted good handwriting. We still gasp at feats of linear memory, Lu Chao reciting 67,890 digits of π or Kim Peek's recall of 12,000 books (60 gigabytes)—even though pathetic compared to the Internet of 10 exabytes (double that in five years).

But the Internet is amazing not for storage (or math) but for connections. Going from footnotes to hypertext to search engines dramatically opens doors for evidence-based thinking, modeling, and collaboration. It transforms itself from mere text to Google Goggles for places and Picasa for faces.

But still it can't do things that Einstein and Curie could. Primate brains changed dramatically from early apes at 400 cc to *Habilis* at 750 cc to Neanderthal at 1,500 cc.

"How did *that* change the way you think?" and "For what purpose?" How will we think to rebuild the ozone after the next nearby supernova? Or nudge the next earth-targeted asteroid? Or contain a pandemic in our dense and well-mixed population? And how will we prepare for those rare events by solving today's fuel, food, psychological, and poverty problems, which prevent 6.7 billion brains from achieving our potential? The answer is blowin' in the Internet wind.

Replacing Experience with Facsimile

Eric Fischl and April Gornik

Visual artists

We might rephrase the question as "How has the Internet changed the way you see?"

For the visual artist, seeing is essential to thought. It organizes information and how we develop thoughts and feelings. It's how we connect.

So, how has the Internet changed us visually? The changes are subtle yet profound. They did not start with the computer. The changes began with the camera and other film-based media, and the Internet has had an exponential effect on that change.

The result is a leveling of visual information, whereby it all assumes the same characteristics. One loss is a sense of scale. Another is a loss of differentiation between materials and the process of making. All visual information "looks" the same, with film/photography being the common denominator.

Art objects contain a dynamism based on scale and physicality that produces a somatic response in the viewer. The powerful visual experience of art locates the viewer very precisely as an integrated self within the artist's vision. With the flattening of visual information and the randomness of size inherent in reproduction, the significance of scale is eroded. Visual information becomes based on image alone. Experience is replaced with facsimile.

As admittedly useful as the Internet is, easy access to images of everything and anything creates a false illusion of knowledge and experience. The world pictured as pictures does not deliver the experience of art seen and experienced physically. It is possible for an art-experienced person to "translate" what is seen online, but the experience is necessarily remote.

As John Berger pointed out in his 1978 essay "The Uses of Photography," the nature of photography is a memory device that allows us to forget. Perhaps something similar can be said about the Internet. In terms of art, the Internet expands the network of reproduction that replaces the way we "know" something. It replaces experience with facsimile.

Outsourcing the Mind

Gerd Gigerenzer

Psychologist; director of the Center for Adaptive Behavior and Cognition at the Max Planck Institute for Human Development, Berlin; author, Gut Feelings

When I came to the Center for Advanced Study in Palo Alto in the fall of 1989, I peered into my new cabinlike office. What struck me was the complete absence of technology. No telephone, e-mail, or other communication facilitators. Nothing could interrupt my thoughts. Technology could be accessed outside the offices whenever one wished, but it was not allowed to enter. This protective belt was there to make sure that scholars had time to think, and to think deeply.

In the meantime, though, the center, like other institutions, has surrendered to technology. Today people's minds are in a state of constant alert, waiting for the next e-mail, the next SMS, as if these will deliver the final, earth-shattering insight. I find it surprising that scholars in the "thinking profession" would so easily let their attention be controlled from the outside, minute by minute, just like letting a cell phone interrupt a good conversation. Were messages to pop up on my screen every second, I would not be able to think straight. Maintaining the center's spirit, I check my e-mail only once a day and keep my cell phone switched off when I'm not making a call. An hour or two without interruption is heaven for me.

But the Internet can be used in an active rather than a reactive way—that is, by not letting it determine how long we can think and when we have to stop. So the question is, Does an active use of the Internet change our way of thinking? I believe so. The Internet shifts our cognitive functions from searching for information inside the mind toward searching outside the mind. But it is not the first technology to do so.

Consider the invention that changed human mental life more than anything else: writing, and subsequently the printing press. Writing made analysis possible; it allowed us to compare texts, which is difficult in an oral tradition. Writing also made exactitude possible, as in higher-order arithmetic—without any written form, these mental skills quickly meet their limits. But writing makes long-term memory less important than it once was, and schools have largely replaced the art of memorization by training in reading and writing.

Most of us can no longer memorize hour-long folktales and songs, as in an oral tradition. The average modern mind has a poorly trained long-term memory, forgets rather quickly, and searches for information more often in outside sources, such as books, rather than from inside memory. The Internet has amplified this trend of shifting knowledge from the inside to the outside and taught us new strategies for finding what we want by using search machines.

This is not to say that before writing, the printing press, and the Internet our minds did not have the ability to retrieve information from outside sources. But these sources were other people, and the skills were social, such as the art of persuasion and conversation. To retrieve information from Wikipedia, say, social skills are unnecessary.

The Internet is essentially a huge storage room of information. We are in the process of outsourcing information storage and retrieval from mind to computer, just as many of us have already outsourced doing mental arithmetic to the pocket calculator. We may lose some skills in this process, such as the ability to concentrate over an extended period of time and the ability to store large amounts of information in long-term memory, but the Internet is also teaching us new skills for accessing information.

It is important to realize that mentality and technology are one extended system. The Internet is a kind of collective memory, to which our minds will adapt until a new technology eventually replaces it. Then we will begin outsourcing other cognitive abilities and—it is to be hoped—learning new ones.

A Prehistorian's Perspective

Timothy Taylor

Archaeologist, University of Bradford, United Kingdom; author, The Artificial Ape: How Technology Changed the Course of Human Evolution

I do not think the Internet has significantly changed the way we think: It was designed for people like me, by people like me, most of them English-speakers. Fundamentally reflecting Western, rationalist, objective, data-organizing drives, the Internet simply enhances my ability to think in familiar ways, letting me work longer, more often, with better focus, free from the social tyranny of the library and the uncertainty of the mails. The Internet has changed what I think, however—most notably about where the human race is now headed. From a prehistorian's perspective, I judge that we have been returned to a point last occupied at the time of our evolutionary origin.

When the first stone tool was chipped more than 2 million years ago, it signaled a new way of being. The ancestral community learned to make flint axes, and those first artificial objects, in turn, critically framed a shared, reflective consciousness that began to express itself in language. An axe could be both made and said, used and asked for. The invention of technology brought the earliest unitary template for human thought into being. It can even be argued that it essentially created us as characteristically human.

What happened next is well known: Technology accelerated adaptation. The original ancestral human culture spread out across continents and morphed into cultures, plural—myriad ways of being. While isolated groups unconsciously drifted into ever greater idiosyncrasy, those who found themselves in competition for the same resources consciously strove to differentiate themselves from their neighbors. This ever deepening cultural specificity facilitated the dehumanization of enemies that successful warfare, driven by jealously guarded technological innovation, required.

Then reunification began, starting 5,000 years ago, with the development of writing—a technology that allowed the transcription of difference. War was not over, but alien thoughts did begin to be translated, at first very approximately, across the boundaries of local incomprehension. The mature Internet marks the completion of this process and thus the reemergence of a fully contiguous human cultural landscape. We now have the same capacity for being united under a common language and shared technology that our earliest human ancestors had.

So in a crucial sense we are back at the beginning, returned into the presence of a shared template for human thought. From now on, there are vanishingly few excuses for remaining ignorant of objective scientific facts, and ever thinner grounds for cultivating hatred through willful failure to recognize our shared humanity. Respecting difference has its limits, however: The fact of our knowing that there is a humanity to share means we must increasingly work toward agreeing on common moral standards. The Internet means that there is nowhere to hide and no way to shirk responsibility when the whole tribe makes informed decisions (as it now must) about its shared future.

The Fourth Phase of *Homo sapiens*

Scott Atran

Anthropologist, National Center for Scientific Research, Paris; author, In Gods We Trust

I received this year's *Edge* question while in Damascus, shuttling messages from Jerusalem aimed at probing possibilities for peace. And I got to thinking about how my thinking on world peace and transnational violence has been shaped by the Internet, and how the advent of the Internet has framed my view of human history and destiny.

I'm aware that I'm living on the cusp of perhaps the third great tipping point in human history, and that this is an awesome and lucky thing to experience.

First, I imagine myself with a small band moving out of Africa into the Fertile Crescent around 60,000 years ago, when humans mastered language and began to conquer the globe. More than half a million years ago, the Neanderthal and human branches of evolution began to split from our common ancestor *Homo erectus* (or perhaps *Homo ergaster*). Neanderthals, like *H. erectus* before, spread out of Africa and across Eurasia. But our ancestors, who acquired fully human body structures about 200,000 years ago, remained stuck in the savanna grasslands and scrub of first eastern and then southern Africa. Recent archaeological and DNA analyses suggest that our species may have tottered on the verge of extinction as recently as 70,000 years ago, dwindling to fewer than 2,000 souls. Then, in an almost miraculous change of fortune about 60,000 to 50,000 years ago, one or a few human bands moved out of Africa for good.

This beginning of human wanderlust was likely stirred by global cooling and the attendant parching of the African grasslands, which led to loss of game and grain. But there is also the strong possibility, based on circumstantial evidence relating to a "cultural explosion" of human artifacts and technologies, that a mutation rewired the brain for computational efficiency. This rewiring allowed for recursion (embedding whole bundles of perceptions and thought within other bundles of perceptions and thoughts), which is an essential property of both human language (syntactic structures) and mind-reading skills (or Theory of Mind, the ability to infer other people's thoughts and perceptions: "I know that she knows that I know that he knows that . . . ," etc.).

Language and mind reading, in turn, became critical to development of peculiarly human forms of thinking and communication, including planning and cooperation among anonymous strangers, imagining plausible versus fictitious pasts and futures, the counterfactuals of reason, and the supernaturals of religion. Together, language and mind reading generated both self-awareness and awareness of others. Other animals may have beliefs, but they don't know they have them. Once humans could entertain and communicate imaginary worlds and beliefs about beliefs, they could break apart and recombine representations of the material and social world at will, with or without regard to immediate or future biological needs.

Human societies, the great French anthropologist Claude Lévi-Strauss argued, divide into "cold" and "hot" cultures. For most of the time that humans have walked the earth, there were only preliterate, "cold" societies, whose people conceived of nature and social time as eternally static or entirely cyclical: That is, the current order was conceived as a projection of an order that had existed since mythical times. The interpretation of the origins of the world and the development of society was rendered in mythological terms. Every element of the knowable universe would be connected in kaleidoscopic fashion to every other element in memorable stories, however arbitrary or fantastic, that could be passed down orally from generation to generation.

A typical mythic account of the world might "explain" how nomadic patterns of residence and seasonal movement emanated from patterns perceived in the stars; how star patterns, in turn, got their shapes from the wild animals around; and how people were made to organize themselves into larger totemic societies, dividing tasks and duties according to the "natural order."

So I imagine myself in ancient Mesopotamia, trying to kick myself out of this cold cycle, as human history began to heat up at the dawn of writing. I try to conjure up in my mind how the seemingly unchanging and cyclical world of oral memory and myth, of frozen and eternal history, could almost all of a sudden, after tens of thousands of years of near stasis, flame forward along the Eurasian Silk Road into civilizations and world commerce, universal religions and government by law, armies and the accumulated knowledge that would one day become science.

Direct reciprocity—"I'll scratch your back and you scratch mine"—works well within small bands or neighborhoods, where people know one another and it would be hard to get away with cheating customers. But as societies become larger and more complex, transactions increasingly involve indirect forms of reciprocity: promises between strangers of delivery after payment, or payment after delivery. Roads, writing, money, contracts, and laws—the channels of communication and exchange that make state-level societies viable—greatly increase prospects for variety, reliability, and accountability in indirect transactions. As groups expand in size, exploiting a widening range of ecological habitats, an increasing division of productive and cognitive labor becomes both possible and preferable.

By the time of Jesus Christ, two millennia ago, four great neighboring polities spanned Eurasia's middle latitudes: the Roman Empire; the Parthian Empire, centered in Persia and Mesopotamia; the Kushan Empire of Central Asia and Northern India; and the Han Empire of China and Korea. The Kushan Empire had diplomatic links with the other three, and all four were linked by a network of trade routes known to posterity as the Silk Road. It's along the Silk Road that Eurasia's three universalist moral religions—Judaism, Zoroastrianism, and Hinduism—interacted and mutated from their respective territorial and tribal origins into the three proselytizing, globalizing religions that today vie for the soul of humanity: Christianity, Islam, and Buddhism.

The three globalizing religions created two new concepts in human thought: individual free choice and collective humanity. People not born into these religions could, in principle, choose to belong (or remain outside), without regard to ethnicity, tribe, or territory. The mission of these religions was to extend moral salvation to all peoples, not just to a "chosen people" that would light the way for others.

Secularized by the European Enlightenment, the great quasi-religious isms of modern history—colonialism, socialism, anarchism, fascism, communism, and democratic liberalism—harnessed industry and science to continue on a global scale the human imperative of "cooperate to compete."

Today I see myself riding on the information highway of cyberspace as if I were on a light beam, casting off such previous human technologies and relationships as books and nation states. If people could fly like Superman, they wouldn't need cars or elevators—and if they can electronically surf for knowledge and relationships, then physical libraries and borders become irrelevant.

I try to imagine what the world will be like with social relationships unbounded by space, and the spiraling fusion of memory and knowledge in a global social brain that anyone can access. Future generations will be able to bind with their ancestors in various ways, because they can see and hear them as they actually were and not just in isolated texts, paintings, and photographs. And the multiple pathways and redundancies in knowledge networks will enable even the simpleminded to approach the creations of genius.

Truth be told, I can no more foresee the actual forms of knowledge, technology, and society that are likely to result than an ancient Bushman or Sumerian could foresee how people could split the atom, traipse on the moon, crack the genetic code, or meet for life in cyberspace. (And anyone who says he can is just blowing smoke in your face.)

But I am reasonably sure that whatever new forms arise, they will have to accommodate fundamental aspects of human nature that have hardly changed since the Stone Age: love, hate, jealousy, guilt, contempt, pride, loyalty, friendship, rivalry, the thrill of risk and adventure, the joy of accomplishment and victory, the desire for esteem and glory, the search for pattern and cause in everything that touches and interests us, and the inescapable need to fashion ideas and relationships sufficiently powerful to deny our nothingness in the random profusion of the universe.

As for future forms of human governance, I see as equally likely (as things look now) the chance that political freedom and diversity, or a brave new world of dumbing homogeneity and deadening control by consensus, will prevail or perhaps alternate in increasingly destructive cycles. For the Internet is currently both the oxygen of a truly open society and of spectacular transnational terrorism.

Here are two snippets that illustrate this duality:

"On the Internet, nobody knows you're a dog," says the cunning canine in Peter Steiner's 1993 *New Yorker* cartoon; and on the Internet, any two communicators can believe they are the world.

"The media is [*sic*] coming!" Skyped the Lashkar-e-Taiba handler to the killers for God at the Taj Hotel in Mumbai, signaling to them that now was the best timing for their martyrdom.

Around the Shi'ite holiday of Ashura—December 28, 2009—I received an e-mail from a friend in Tehran who said how helpless he felt to stop the merciless beating of a young woman by government thugs, but he went on to say, "We will win this thing if the West does nothing but help us keep the lines of communication open with satellite Internet." The same day, I saw the Facebook communications of the Christmas plane bomber and the army psychiatrist who shot up Fort Hood—both of them, along with many others, self-bound into a virtual community whose Internet imams spin Web dreams of glory in exchange for real and bloody sacrifice.

"I imagine how the great jihad will take place, how the Muslims will win, *insha Allah* [God willing], and rule the whole world, and establish the greatest empire once again!!!" reads one post from "farouk 1986," the angel-faced, British-educated engineering student and son of a prominent Nigerian banker who attempted to blow up Northwest flight 253 out of Amsterdam as it was about to land in Detroit. "Happiness is martyrdom" can be as emotionally contagious to a lonely boy on the Internet as "Yes, we can." That is a psychologically stunning and socially far-reaching development that scientists have hardly begun to explore.

And so, as a result of the advent of the Internet, I spend most of my time these days trying to think how, with the aid of the Internet, to keep farouk 1986 and friends from blowing people to kingdom come.

The Collapse of Cultures

Human rights constitutes a pillar of one global political culture, originally centered on the Americas and Europe, and is a growing part of a massive, Internet-driven global political awakening. The decidedly nonsecular jihad is another key mover in this transnational political awakening: thoroughly modern and innovative, despite its atavistic cultural references. Its appeal, to youth especially, lies in its promise of moral simplicity, a harmonious and egalitarian community (at least for men) whose extent is limitless, and the call to passion and action on humanity's behalf. It is a twisting of the tenets of human rights, the granting to each individual the "natural right" of sovereignty. It claims a moral duty to annihilate any opposition to the coming of true justice, and gives the righteous the prerogative to kill. The end justifies the means; no sacrifice of individuals is too costly for progress toward the final good.

Many made giddy by globalization—the ever faster and deeper integration of individuals, corporations, markets, nations, technologies, and knowledge—believe that a connected world inexorably shrinks differences and divisions, making everyone safer and more secure in one great big happy family. If only it were not for people's premodern parochial biases: religions, ethnicities, native languages, nations, borders, trade barriers, historical chips on the shoulder.

This sentiment is especially common among scientists (me included) and the deacons of Davos, wealthy and powerful globetrotters who schmooze with one another in airport VIP clubs, three-star restaurants, and five-star hotels and feel that pleasant buzz of camaraderie over wine or martinis at the end of the day. I don't reject this world; I sometimes embrace it.

But my field experience and experiments in a variety of cultural settings lead me to believe that an awful lot of people on this planet respond to global connectivity very differently than does the power elite. While economic globalization has steamrolled or left aside large chunks of humankind, political globalization actively engages people of all societies and walks of life—even the global economy's driftwood: refugees, migrants, marginals, and those most frustrated in their aspirations.

For there is, together with a flat and fluid world, a more tribal, fragmented, and divisive world, as people unmoored from millennial traditions and cultures flail about in search of a social identity that is at once individual and intimate but with a greater sense of purpose and possibility of survival than a human, or humankind, alone.

Ever since the collapse of the Soviet Union, which shattered the briefly timeless illusion of a stable bipolar world, and for the first time in history, most of humanity is politically engaged. Many, especially the young, are becoming increasingly independent, yet interactive in the search for respect and meaning in life, in their visions of economic advancement and environmental awareness. These youth form their identities in terms of global political cultures through exposure to the media.

Even justice for the blistered legacies of imperialism and colonialism is now more a struggle over contemporary construction of cultural identity, and how the media should represent the past, than over righting the actual wrongs that were perpetrated. Global political cultures arise horizontally, among peers with different histories, rather than vertically as before, in traditions passed from generation to generation. Jihad offers the pride of great achievements for the underachieving: brave new hearts for an outworn and overstretched world.

Traditionally, politics and religion were closely connected to ethnicity and territory, and in more recent times to nations and cultural areas (or "civilizations"). No longer. Religion and politics are becoming increasingly detached from their cultures of origin, not so much because of the movement of peoples (only about 3 percent of the world's population migrates, notes French political scientist Olivier Roy) but because of the worldwide traffic of media-friendly information and ideas. Thus, contrary to those who see global conflicts along long-standing "fault lines" and a "clash of civilizations," these conflicts represent a collapse of traditional territorial cultures, not their resurgence. The crisis is most likely to be resolved, I believe, in cyberspace. To what end, I cannot tell but can only hope.

Transience Is Now Permanence

Douglas Coupland

Writer, artist, designer; author, Generation A

The Internet has made me very casual, with a level of omniscience that was unthinkable a decade ago. I now wonder if God gets bored knowing the answer to everything.

The Internet forces me, as a creator, to figure out who I really am and what is unique to me—or to anyone else, for that matter. I like this.

The Internet forces me to come to grips with the knowledge that my mother has visited many truly frightening places online that I'll never know about—and certainly don't want to know about. I no longer believe in a certain sort of naïveté.

The Internet toys with my sense of permanence. Every tiny transient moment now lasts forever: homework, e-mails, JPEGs, sex acts . . . we all know the list. Yesterday I looked up a discontinued brand of Campbell's Soup called Noodles and Ground Beef and was taken (via Google Books) to page 37 of the February 1976 issue of *Ebony* magazine, to a recipe for Beefy Tomato Burger Soup that incorporated a can of the aforementioned soup. You'd have thought something that ephemeral would have evaded Google's reach, but no. Transience is now permanence. At the same time, things that were supposed to be around forever (newspapers) are now transient. This is an astonishing inversion of time perception that I've yet to fully absorb. Its long-term effect on me is to heighten my worry about the fate of the middle classes (doomed) as well as to make me wonder about the future of homogeneous bourgeois thinking (also doomed, as we turn into one great big college town populated entirely by eccentrics—a great big Austin, Texas).

The Internet forces me to renegotiate my relationship to the celebrity dimension of pop culture. There are too many celebrities now, and they all cancel each other out (every fifteen minutes), so there aren't megastars, like there used to be. You might as well be eccentric yourself.

The Internet gives me hope that in the future everyone will wear Halloween costumes 365 days a year.

The Internet has certainly demystified my sense of geography and travel. On Google Maps, I've explored remote Antarctic valleys as well as Robert Smithson's sculptural earthwork "Spiral Jetty." And we've all taken BlackBerrys everywhere. In so many ways, anywhere is basically as good as anywhere else—so let's hope you ended up somewhere with a nice climate and pleasant scenery when the music stopped in the fall of 2008.

Speaking of music, the Internet has made me much more engaged with musical culture than I might have hoped for when coming of age in the 1970s. It used to be that a person's musical taste was frozen around the age of twenty-three. Once this happened, a person (usually a guy) spent the rest of his life worshipping stacks of lovingly maintained 33 rpm vinyl. Nowadays the curation of an individual's personal taste never ends. People don't ask, "Have you heard the new [whatever]?" Instead it's "What have you found lately?" It's friendlier and allows for communication between people of all ages.

A Return to the Scarlet-Letter Savanna

Jesse Bering

Psychologist; director, Institute of Cognition and Culture, Queens University, Belfast; columnist, Scientific American *("Bering in Mind"); author,* The Belief Instinct: The Psychology of Souls, Destiny, and the Meaning of Life

Only 10,000 years ago, our *Homo sapiens* ancestors were still living in close-knit societies about the size of which today would barely fill a large lecture hall in a state university. What today might be seen as an embarrassing faux pas could have been the end of the line for you back then. At least, it could have been the end of the line for your reproductive success, since an irreversibly spoiled reputation in such a small group could have meant a surefire death for your genes.

Just imagine the very worst thing you've ever done: the most vile, scandalous, and vulgar. Now imagine all the details of this incident tattooed on your forehead. This scenario is much like what our ancestors would have encountered if their impulsive, hedonistic, and self-centered drives weren't kept in check by their more recently evolved prudent inhibitions. And this was especially the case, of course, under conditions in which others were watching them, perhaps without their realizing it. Our ancestors, if their ancient, selfish drives overpowered them, couldn't simply pull up stakes and move to a new town. Rather, since they were more or less dependent on those with whom they shared a few hundred square kilometers, cutting off all connections wasn't a viable option. And hiding their identities behind a mantle of anonymity wasn't really doable, either, since they couldn't exactly be a nameless face. The closest our ancestors had to anonymity was the cover of night. Thus in the ancestral past, being good, being moral by short-circuiting our species' evolved selfish desires, was even more a matter of life and death than it is today. It was a scarlet-letter savanna.

Yet curiously, for all its technological sophistication and seeming advances, the Internet has heralded something of a return to this scarlet-letter-savanna environment and in many ways has brought our species back to its original social roots.

After a long historical period during which people may have been able to emigrate to new social groups and start over if they had spoiled their reputations, the present media age more accurately reflects the conditions faced by our ancestors. With newspapers, telephones, cameras, television, and especially the Internet at our disposal, personal details about medical problems, spending activities, criminal and financial history, and divorce records (to name just a few matters potentially costly to our reputations) are not only permanently archived but also can be distributed in microseconds to, literally, millions of other people. The old adage "Wherever you go, there you are" takes on new meaning in light of the evolution of information technology. From background checks to matchmaking services to anonymous Website browsing to piracy and identity theft, from Googling others (and ourselves) to flaming bad professors (e.g., www.ratemyprofessor.com) and stingy customers (e.g., www.bitterwaitress.com), the Internet is simply ancient social psychology meeting new information technology.

Take Love

Helen Fisher

Research professor, Department of Anthropology, Rutgers University; author, Why We Love

For me, the Internet is a return to yesteryear; it simply allows me (and all the rest of us) to think and behave in ways for which we were built long, long ago. Take love. For millions of years, our forebears traveled in little hunting-and-gathering bands. Some twenty-five individuals lived together, day and night; some ten to twelve were children and adolescents. But everyone knew just about everyone else in a neighborhood of several hundred square miles. They got together, too. Annually, in the dry season, bands congregated at the permanent waters that dotted eastern and southern Africa. Here as many as five hundred men, women, and children would mingle, chat, dine, dance, perhaps even worship together. And although a pubescent girl who saw a cute boy at the next campfire might not know him personally, her mother probably knew his aunt, or her older brother had hunted with his cousin. All were part of the same broad social web.

Moreover, in the ever present gossip circles a young girl could easily collect data on a potential suitor's hunting skills, even on whether he was amusing, kind, smart. We think it's natural to court a totally unknown person in a bar or club. But it's far more natural to know a few basic things about an individual before meeting him or her. Internet dating sites, chat rooms, and social networking sites provide these details, enabling the modern human brain to pursue more comfortably its ancestral mating dance.

Then there's the issue of privacy. Some are mystified by the way others, particularly the young, so frivolously reveal their intimate lives on Facebook or Twitter, in e-mails, and via other Internet billboards. This odd human habit has even spilled into our streets and other public places. How many times have you had to listen to people nonchalantly blare out their problems on cell phones while you sat on a train or bus? Yet for millions of years our forebears had almost no privacy. With the Internet, we are returning to this practice of shared community.

So for me, the Internet has only magnified—on a grand scale—what I already knew about human nature. Sure, with the Net, I more easily and rapidly acquire information than in the old days. I can more easily sustain connections with colleagues, friends, and family. And sometimes I find it easier to express complex or difficult feelings via e-mail than in person or on the phone. But my writing isn't any better . . . or worse. My perspectives haven't broadened . . . or narrowed. My values haven't altered. I have just as much data to organize. My energy level is just the same. My workload has probably increased. And colleagues want what they want from me even faster. My daily habits have changed—moderately.

But the way I think? I don't think any harder, faster, longer, or more effectively than I did before I bought my first computer, in 1985. In fact, the rise of the Internet only reminds me of how little any of us have changed since the modern human brain evolved more than 35,000 years ago. We are still the same warlike, peace-loving, curious, gregarious, proud, romantic, opportunistic, and naïve creatures we were before the Internet—indeed, before the automobile, the radio, the Civil War, or the ancient Sumerians. We still have the same brain our forebears had as they stalked woolly mammoths and mastodons; we still chat and warm our hands where they once camped—on land that is now London, Beijing, New York. With the Internet, we just have a much louder megaphone with which to scream who we really are.

Internet Mating Strategies

David M. Buss

Professor of psychology, University of Texas, Austin; coauthor (with Cindy M. Meston), Why Women Have Sex

The ancient strategies of human mating are implemented in novel ways on the Internet. Humans evolved in small groups with available mates limited to a few dozen possibilities. The Web provides unprecedented and tantalizing access to thousands, or millions. The stigma of traditional dating services, once the refuge of the lonely and forlorn, has disappeared in the digital world of modern mating.

The bounty of mating opportunities in today's computational sphere yields some tangible benefits. It allows people to secure better mating fits—access to that special someone who shares your unique interests in underground rock bands, obscure novelists, or unheard-of foreign movies. It can abbreviate search costs, eliminating the nonstarters without forcing you to slog through the cumbersome dating maze. The Internet affords practice (the stuttering and shy in person can be eloquently bold on the keyboard). Because of the surfeit of opportunity, the Internet may yield good bargains on the mating market, a maximization of one's mate value, or access to the otherwise unattainable. It allows some to luxuriate in sexual adventures unimaginable in the small-group living of our distant past.

Humans are loath to settle when better prospects entice. The abundance of mating opportunities sometimes produces paralyzing indecision. A more exciting encounter, a more attractive partner, a true soul mate might be a few clicks away. The World Wide Web may reduce commitment to a "one and only," because opportunities for promising others seem so plentiful. It can cloak sexual deception.

Are the personal descriptions accurate? Are images Photoshopped? It opens new avenues for exploitation. Sexual predators hone their tactics on the unwary, the innocent, or those open to adventure. At the same time, computer-savvy victims countermand those maneuvers, manipulate marauders, and reduce their vulnerability to predation in a never-ending arms race.

In most ways, though, the Internet has not altered how we *think* about mating. Nor has it changed our underlying sexual psychology. Men continue to value physical appearance. Women continue to value ambition, status, and financial prospects. Both sexes continue to trade up when they can and cut losses when they can't. Sexual economics remain; only the format has changed. Hunter-gatherer market exchanges of sex and meat have been replaced with Internet markets of sugar babies and sugar daddies. The mating and dating sites most successful are those that exploit our ancient mating psychology. Evolved mechanisms of mind now can be played out in the global, semianonymous modern world of interconnectivity. The eternal quest for love, spirituality, or sexual union may evaporate in the clouds of cyberspace. But then again, glory in affairs of the heart has always been fleeting.

Internet Society

Robert R. Provine

Psychologist and neuroscientist, University of Maryland; author, Laughter: A Scientific Investigation

At the end of my college lectures, students immediately flip open their cell phones, checking for calls and texts. In the cafeteria, I observe students standing in queues, texting, neglecting fellow students two feet away. Late one afternoon, I noticed six students wandering up and down a hallway while using cell phones, somehow avoiding collision, like ships cruising in the night, lost in a fog of conversation—or like creatures from *The Night of the Living Dead.* A student reported e-mailing on a Saturday night "computer date" without leaving her room. Paradoxically, these students were both socially engaged and socially isolated.

My first encounter with people who were using unseen phone headsets was startling; they walked through an airline terminal apparently engaging in soliloquies or responding to hallucinated voices. More is involved than the displacement of snail mail by e-mail, a topic of past decades. Face-to-face encounters are being displaced by relations with a remote, disembodied conversant somewhere in cyberspace. These experiences forced a rethinking of my views about communication—technological and biological, ancient and modern—and prompted research projects examining the emotional impact, novelty, and evolution of social media.

The gold standard for interpersonal communication is face-to-face conversation, in which you can both see and hear your conversant. In several studies, I contrasted this ancestral audiovisual medium with cell phone use, in which you hear but do not see your conversant, and texting, in which you neither see nor hear your conversant.

The telephone, whether cell or land line, provides a purely auditory medium that transmits two-way vocal information, including the prosodic (affective) component of speech. Although it filters the visual signals of gestures, tears, smiles, and other facial expressions, the purely auditory medium of the telephone is itself socially and emotionally potent, generating smiles and laughter in remote individuals—a point we confirmed by observation of 1,000 solitary people in public places. Unless using a cell phone, isolated people are essentially smileless, laughless, and speechless. (We confirmed the obvious because the obvious is sometimes wrong.) Constant, emotionally rewarding vocal contact with select, distant conversants is a significant contributor to the worldwide commercial success of cell phones. Radio comedy and drama further demonstrate the power of a purely auditory medium, even when directed one-way from performer to audience. It occurred to me that the ability to contact unseen conversants is a basic property of the auditory sense; it's as old as our species and occurs every time we speak with someone in the dark or not in our line of sight. Phones become important when people are beyond shouting distance.

Conversations between deaf signers provided a medium in which individuals could see but not hear their conversant. With my collaborator Karen Emmorey, I explored the emotional communication between them. We observed vocal laughter and associated social variables in conversations between deaf signers who were using American Sign Language. Despite their inability to hear their conversational partner, they laughed at the same places in the stream of signed speech, at similar material, and showed the same gender patterns of laughter as hearing individuals during vocal conversations. An emotionally rich dialog thus can be conducted with an exclusively visual medium that filters auditory signals and passes only visual ones. Less nuanced visual communication is ancient, and used when communicating beyond vocal range, via such signals as gestures, flags, lights, mirrors, or smoke.

Text messaging, however—whether meaty e-mails or telegraphic tweets—involves conversants who can neither see nor hear each other and are not interacting in real time. My research team examined emotional communication online by analyzing the placement of 1,000 emoticons in Website text messages. Emoticons seldom interrupted phrases. For example, you may text, "You are going where on vacation? Lol" but not "You are—lol—going where on vacation?" Technophiles writing about text messaging sometimes justify emoticon use as a response to the "narrowing of bandwidth" characteristic of text messaging, ignoring that text viewed on a computer monitor or cell phone is essentially identical to that of the printed page. I suspect that emoticon use is a likely symptom of the limited literary prowess of texters. Know what I mean? Lol. Readers seeking the literary subtleties of irony, paradox, sarcasm, or sweet sorrow are unlikely to find it in text messages.

The basic cell phone has morphed into a powerful mobile multimedia communication device and computer terminal that is a major driver of Internet society. It gives immediate, constant contact with select, distant conversants; can tell you where you are, where you should go next, and how to get there; provides diversions while waiting; and can document your journey with text, snaps, and video images. For some, this is enhanced reality, but it comes at the price of the here and now. Whatever your opinion and level of engagement, the cell phone and related Internet devices are profound social prostheses—almost brain implants—that have changed our lives and culture.

Don't Ring Me

Aubrey de Grey

Gerontologist; chief science officer, SENS Foundation; author (with Michael Rae), Ending Aging

The Net changes the way I think in a bunch of ways that apply to more or less everyone, and especially to *Edge* question respondents, but there's one effect it has on me that is probably rarer. And it's not a change but an avoidance of a change.

Before I switched to biology, I was a computer scientist; I have been using e-mail regularly since I was a student in the early 1980s. And I like e-mail—a lot. E-mail lets you think before you speak, on those frequent occasions when doing so would be a good idea. E-mail waits patiently for you to read it, and the sender isn't offended if you reply a few hours or even a day after you get it. E-mail lets you speak in real sentences when you want to—and not in real sentences when you don't.

What might I be thinking of that so offensively lacks those qualities? No, not face-to-face interaction: I am as gregarious as anyone. Not snail mail, either—though I certainly use that medium far more rarely now than I did a decade or two ago. No, the object of my distaste is the greatest curse of the twenty-first century, the cell phone.

It would take more words than we have been allowed for these pieces to do full justice to my loathing of the cell phone, so I won't try. But you can probably guess that it doesn't stop at the irritation caused when someone's phone goes off in the middle of a lecture. A lot of it is the sheer rudeness that cell phones force their owners to commit, in situations where no such problem would otherwise exist: to wit, abruptly suspending a face-to-face conversation to take a call, or summarily telling someone to call back because the person you're talking to is more important. But most of it is the contrast with the civilized, relaxed, entirely adequate form of communication that I so prefer: e-mail.

Yes, yes, you're going to protest that one can always turn one's phone off. That's nonsense. If there's one thing worse than being rung when you don't want to be, it's having someone ask you to ring them, doing so, and then getting their voice mail. Hello? If I wanted to tell you something without hearing your immediate response, I'd have sent you an e-mail (as I wanted to do in the first place).

As the cell phone has become increasingly ubiquitous, I have come under increasing pressure to conform. So far I have resisted, and there is every sign that I shall continue to do so. How? Simply because I'm very well behaved with e-mail. With the few percent of e-mails I receive to which I want to take time to compose a reply, I take that time—but for the great majority, I'm fast. Really fast. It's the best of both worlds: negligible slowdown in communication without the loss of that resource so rare and valuable to the busy high achiever, occasional but reliable solitude. And also without the other drawbacks I've mentioned. Put simply, I'm easy enough to interact with using e-mail. If the Internet didn't exist, or if it weren't so ubiquitous, I'd have been forced long ago to submit to the tyranny of the cell phone and I would be an altogether less nice person to know.

A Thousand Hours a Year

Simon Baron-Cohen

Psychologist, Autism Research Centre, Cambridge University; author, Autism and Asperger Syndrome: The Facts

Possibly like you, all my e-mail goes into my Sent mailbox, just sitting there in case I want to check back on what I said to whom years ago. So what a surprise to see that I send approximately 18,250 e-mails each year (roughly 50 a day). Assuming three minutes per e-mail (let's face it, I can't afford to spend too long thinking about what I want to say), that's about 1,000 hours a year on e-mail alone. I've been on e-mail since the early nineties. Was that time well spent?

The answer is both yes and no. Yes, I have been able to keep in touch with family, friends, and colleagues in far-flung corners of the planet with ease, and I have managed to pull off projects with teams spread across different cities in time scales that previously would have been unthinkable. All this feeds my continued use of e-mail. But whereas these undoubted benefits are the reasons why I continue to e-mail, e-mailing is not without its cost. Most important, as my analysis shows, e-mail eats my time, just as it likely eats yours. And unlike Darwin's famous 15,000 letters (penned with thought, and now the subject of the Darwin Correspondence Project in my university library in Cambridge), three-minute e-mail exchanges do not deliver communication with any depth and, as such, are not intellectually valuable in their own right.

We all recognize that e-mail has its addictive side. Each time a message arrives, there's the chance it might contain something exciting, something new, something special, a new opportunity. Like all effective behavioral reinforcement schedules, the reward is intermittent: Maybe one in a hundred e-mails contains something I really want to know or hear about. That's just enough to keep me checking my inbox, but it means that perhaps only 10 of the 1,000 hours I spent on e-mails this year were rewarded.

Bite-size e-mails also carry another cost. We all know there's no substitute for thinking hard and deep about a problem and how to solve it, or for getting to grips with a new area; such tasks demand long periods of concentrated attention. Persistent, frequent e-mail messages threaten our capacity for the real work. Becoming aware of what e-mail is doing to our allocation of time is the first step to regaining control. As in stifling other potential addictions, we should perhaps attempt to counter the e-mail habit by restricting it to certain times of the day, or by creating e-mail-free zones by turning off Wi-Fi. This year's *Edge* question at least gives me pause to think whether I really want to be spending a thousand hours a year on e-mail at the expense of more valuable activities.

Thinking Like the Internet, Thinking Like Biology

Nigel Goldenfeld

Physicist, University of Illinois at Urbana-Champaign

Although I used the Internet back when it was just ARPANET—and even earlier, as a teenager, using a teletype to log on to a state-of-the-art Honeywell mainframe from my school—I don't believe my way of thinking was changed by the Internet until around 2000. Why not?

Back in my school days, the Internet was linear, predictable, and boring. It never talked back. When I hacked into the computer at MIT, running an early symbolic-manipulator program, something that could do algebra in a painfully inadequate way, I just used the Internet as a perfectly predictable tool. In my day-to-day life as a scientist (the theoretical physicist geek from central casting), I mostly still do.

In 1996, I cofounded a software company that built its products and operated essentially entirely through the Internet; whether this was more efficient than a bricks-and-mortar company is debatable, but the fact was that through this medium fabulously gifted individuals were able to participate—people who never would have dreamed of relocating for such work. But this was still a linear, predictable, and essentially uninteresting use of the Internet.

No, the Internet is changing the way I think because its whole is greater than the sum of its parts—because of its massive connectivity and the resulting emergent phenomena. When I was a child, they said we would be living on the moon, that we would have antigravity jet packs and videophones. They lied about everything but the videophones. Via private blogs, Skype, and a $40 webcam, I can collaborate with my colleagues, write equations on my blackboard, and build networks of thought that stagger me with their effectiveness. My students and I work together so well through the Internet that its always-on library dominates our discussions and helps us find the sharp questions that drive our research and thinking infinitely faster than before.

My day job is to make discoveries through thought, principally by exploiting analogies through acts of intellectual arbitrage. When we find two analogous questions in what were previously perceived to be unrelated fields, one field will invariably be more developed than the other, so there is a scientific opportunity. This is how physicists go hunting. The Internet has become a better tool than the old paper scientific literature because it responds in real time.

To see why this is a big deal for me, consider the following "homework hack." You want to become an instant expert in something that matters to you: maybe a homework assignment, maybe researching a life-threatening disease afflicting someone close to you. You could research it on the Internet using a search engine—but, as you know, you can search but you can't really find. Google gives you unstructured information, and for a young person in a hurry that is simply not good enough. Search engines are a linear, predictable, and essentially uninteresting way to use the Internet.

Instead, try the following hack. Step 1: Make a wiki page on the topic. Step 2: Fill it with complete nonsense. Step 3: Wait a few days. Step 4: Visit the wiki page and harvest the results of what generous and anonymous souls from—well, who cares where they're from or who they are?—have corrected and enhanced in, one presumes, fits of righteous indignation. It really works. I know because I have seen both sides of this transaction. There you have it: the emergence of a truly global, collective entity, something that has arisen from humans plus the Internet. It talks back.

This homework hack is, in reality, little more than the usual pattern of academic discourse but carried out, in science fiction master

William Gibson's memorable phrase, with "one thumb permanently on the fast-forward button." Speed matters, because life is short. The next generation of professional thinkers already have all the right instincts about the infinite library that is their external mind, accessible in real time, and capable of accelerating the already Lamarckian process of evolution in thought and knowledge on time scales that really matter. I'm starting to get it, too.

Roughly 3 billion years ago, microbial life invented the Internet and Lamarckian evolution. For them, the information is stored in molecules and recorded in genes transmitted between consenting microbes by a variety of mechanisms we are still uncovering. Want to know how to become a more virulent microbial pathogen? Download the gene! Want to know how to hotwire a motorcycle? Go to the Website! So much quicker than random trial-and-error evolution, and it works . . . right now! And your children's always-on community of friends, texting lols and other quick messages that really say "I'm here, I'm your friend, let's have a party," is no different than the quorum sensing of microbes, counting their numbers so that they can do something collectively, such as invade a host or grow a fruiting body from a biofilm.

I'm starting to think like the Internet, starting to think like biology. My thinking is better, faster, cheaper, and more evolvable because of the Internet. And so is yours. You just don't know it yet.

The Internet Makes Me Think in the Present Tense

Douglas Rushkoff

Media analyst; documentary writer; author, Life, Inc.: How the World Became a Corporation and How to Take It Back

How does the Internet change the way I think? It puts me in the present tense. It's as if my cognitive resources have been shifted from my hard drive to my RAM. That which is happening right now is valued, and everything in the past or future becomes less relevant.

The Internet pushes us all toward the immediate. The now. Every inquiry is to be answered right away, and every fact or idea is only as fresh as the time it takes to refresh a page. And as a result, speaking for myself, the Internet makes me mean. Resentful. Short-fused. Reactionary.

I feel it when I'm wading through a stack of e-mails, keeping up with an endless Twitter feed, accepting Facebook "friends" from a past I prefer not to remember, or making myself available on the Web to readers to whom I should feel grateful but instead feel obligated. And it's not a matter of what any of these folks might want me to do, but rather when. They want it now.

This is not a bias of the Internet itself but of the way it has changed from an opt-in activity to an always-on condition of my life. The bias of the medium was never toward real-time activity but toward time shifting. UNIX, the operating system of the Net, doesn't work in real time; it sits and waits for human commands. Likewise, early Internet forums and bulletin boards were discussions that users returned to at their convenience. I dropped in on the conversation, then came back the next evening or the next week to see how it had developed. I took the time to consider what I might say—to contemplate someone else's response. An Internet exchange was only as rich as the amount of time I allowed to pass between posts.

Once the Internet changed from a resource at my desk into an appendage chirping from my pocket and vibrating on my thigh, however, the value of depth was replaced by that of immediacy masquerading as relevancy. This is why Google is changing itself from a search engine to a "live" search engine, why e-mail devolved to SMS and blogs devolved to tweets. It's why schoolchildren can no longer engage in linear argument, why narrative structure collapsed into reality TV, why almost no one can engage in meaningful dialog about long-term global issues. It creates an environment in which a few incriminating e-mails between scientists generate more news than our much slower but much more significant climate crisis.

It's as if the relentless demand of networks for me to be everywhere, all the time, was denying me access to the moment in which I am really living. And it is this sense of disconnection—more than distraction, multitasking, or long-distance engagement—that makes the Internet so aggravating.

In some senses, this was the goal of those who developed the computers and networks on which we depend today. Technology visionaries such as Vannevar Bush and James Licklider sought to develop machines that could do our remembering for us. Computers would free us from the tyranny of the past (as well as the horrors of World War II), allowing us to forget everything and devote our minds to solving the problems of today. The information would still be there; it would simply be stored out of body, in a machine.

That might have worked had technological development leaned toward the option of living life disconnected from those machines whenever access to their memory banks was not required. Instead, I feel encouraged to use networks not just to access information but to access other people and grant them access to me—wherever and whenever I happen to be.

This always-on approach to digital technology overwhelms my nervous system rather than expanding it. Likewise, the simultaneity of information streaming toward me prevents parsing or consideration. It becomes a constant flow that must be managed, perpetually.

The nowness of the Internet engenders impulsive, unthinking responses instead of considered ones, and a tendency to think of communications as a way to bark orders or fend off those of others. I want to satisfy the devices chirping and vibrating in my pockets, if only to make them stop. Instead of looking at each digital conversation as an opportunity for depth, I experience them as involuntary triggers of my nervous system. Like my fellow networked humans, I now suffer the physical and emotional stresses previously associated with careers such as air traffic controller and 911 operator.

By surrendering my natural rhythms to the immediacy of my networks, I am optimizing myself and my thinking to my technologies rather than the other way around. I feel as though I am speeding up when I am actually becoming less productive, less thoughtful, and less capable of asserting any agency over the world in which I live. The result is something akin to future shock. Except that in our era, it's more of a present shock.

I try to look at the positive: Our Internet-enabled emphasis on the present may have liberated us from the twentieth century's dangerously compelling ideological narratives. No one—well, hardly anyone—can still be persuaded that brutal means are justified by mythological ends. And people are less likely to believe employers' and corporations' false promises of future rewards for their continued loyalty.

But—for me, anyway—the Internet has not brought greater awareness of what is going on around us. I am not approaching some Zen state of an infinite moment, completely at one with my surroundings, connected to others and aware of myself on any fundamental level. Rather, I am increasingly in a distracted present, where forces on the periphery are magnified and those immediately before me are ignored. My ability to create a plan—much less follow through on it—is undermined by my need to be able to improvise my way through any number of external effects that stand to derail me at any moment. Instead of finding a stable foothold in the here and now, I end up reacting to an ever-present assault of simultaneous impulses and commands.

The Internet tells me I am thinking in real time, when what it really does, increasingly, is take away the real and take away the time.

Social Prosthetic Systems

Stephen M. Kosslyn

Psychologist, dean of social sciences, Harvard University; coauthor (with Robin S. Rosenberg), Fundamentals of Psychology in Context

Other people can help us compensate for our mental and emotional deficiencies, much as a wooden leg can compensate for a physical deficiency. Specifically, other people can extend our intelligence and help us understand and regulate our emotions. I've argued that such relationships can become so close that other people essentially act as extensions of oneself, much as a wooden leg does. When another person helps us in such ways, he or she is participating in what I've called a "social prosthetic system." Such systems do not need to operate face-to-face, and it's clear to me that the Internet is expanding the range of my own social prosthetic systems. It's already an enormous repository of the products of many minds, and the interactive aspects of the evolving Internet are bringing it ever closer to the sort of personal interactions that underlie such systems.

Even in its current state, the Internet has extended my memory, perception, and judgment.

Regarding memory: Once I look up something on the Internet, I don't need to retain all the details for future use—I know where to find that information again and can quickly and easily do so. More generally, the Internet functions as if it were my memory. This function of the Internet is particularly striking when I'm writing; I'm no longer comfortable writing if I'm not connected to the Internet. It's become natural to check facts as I write, taking a minute or two to dip into PubMed, Wikipedia, or the like. When I write with a browser open in the background, it's as though the browser were an extension of myself.

Regarding perception: Sometimes I feel as if the Internet has granted me clairvoyance. I can see things at a distance. I'm particularly struck by the ease of using videos, allowing me to witness a particular event in the news. It's a cliché, but the world really does feel smaller.

Regarding judgment: The Internet has made me smarter in matters small and large. For example, when I'm writing a textbook, it has become second nature to check a dozen definitions of a key term, which helps me distill the essence of its meaning. But more than that, I now regularly compare my views with those of many others. If I have a "new idea," I now quickly look to see whether somebody else has already conceived of it, or something similar—and I then compare what I think to what others have thought. This inevitably hones my own views. Moreover, I use the Internet for sanity checks, trying to gauge whether my emotional reactions to an event are reasonable by quickly comparing them with those of others.

These effects of the Internet have become even more striking since I've begun using a smartphone. I now regularly pull out my phone to check a fact, watch a video, read blogs. Such activities fill the spaces that used to be dead time (such as waiting for somebody to arrive for a lunch meeting).

But that's the upside. The downside is that in those dead periods I often would let my thoughts drift and sometimes would have an unexpected insight or idea. Those opportunities are now fewer and farther between. Like anything else, constant connectivity has posed various tradeoffs; nothing is without a price. But in this case—on balance—it's a small price. I'm a better thinker now than I was before I integrated the Internet into my mental and emotional processing.

Evolving a Global Brain

W. Tecumseh Fitch

Department of Cognitive Biology, University of Vienna; author, The Evolution of Language

When I consider the effect of the Internet on my thought, I keep coming back to the same metaphor. What makes the Internet fundamentally new is the many-to-many topology of connections it allows. Suddenly any two Internet-equipped humans can transfer essentially any information, flexibly and efficiently. We can transfer words, code, equations, music, or video anytime to anyone, essentially for free. We are no longer dependent on publishers or media producers to connect us. This parallels what happened in animal evolution as we evolved complex brains controlling our behavior, partly displacing the basically hormonal, one-to-many systems that came before. So let's consider this new information topology from the long evolutionary viewpoint, by comparing it with the information revolution that occurred during animal evolution over the last half billion years: the evolution of brains.

Our planet has been around for 4.5 billion years, and life appeared very early, almost 4 billion years ago. But for three-quarters of the subsequent period, life was exclusively unicellular, similar to today's bacteria, yeast, or amoebae. The most profound organic revolution, after life itself, was thus the transition to complex multicellular organisms, such as trees, mushrooms, and ourselves.

Consider this transition from the viewpoint of a single-celled organism. An amoeba is a self-sufficient entity, moving, sensing, feeding, and reproducing independent of other cells. For 3 billion years of evolution, our ancestors were all free-living cells like this, independently "doing it for themselves," and they were honed by this long period into tiny organisms more versatile and competent than any cell in our multicellular bodies. Were it capable of scorn, an amoeba would surely scoff at a red blood cell as little more than a stupid bag of protoplasm, barely alive, overdomesticated by the tyranny of multicellular specialization.

Nonetheless, being jacks of all trades, such cells were masters of none. Cooperative multicellularity allowed cells to specialize, mastering the individual tasks of support, feeding, and reproduction. Specialization and division of labor allowed teams of cells to vastly outclass their single-celled ancestors in terms of size, efficiency, and complexity, leading to a whole new class of organisms. But this new organization created its own problems of communication: how to ensure smooth, effective cooperation among all of these independent cells? This quandary directly parallels the origin of societies of specialized humans.

Our bodies have essentially two ways of solving the organizational problems raised by coordinating billions of semi-independent cells. In hormonal systems, master control cells broadcast potent signals all other cells must obey. Steroid hormones such as estrogen or testosterone enter the body's cells, penetrating their nuclei and directly controlling gene expression. The endocrine system is like an immensely powerful dictatorship, issuing sweeping edicts that all must obey.

The other approach involves a novel cell type specialized for information processing: the neuron. While the endocrine approach works fine for plants and fungi, metazoans (multicellular animals) move, sense, and act, requiring a more subtle, neural form of control. From the beginning, neurons were organized into networks: They are team workers collaboratively processing information and reaching group decisions. Only neurons at the final output stage, such as motor neurons, retain direct power over the body. And even motor neurons must act together to produce coordinated movement rather than uncontrolled twitching.

In humans, language provided the beginnings of a communicative organizational system, unifying individuals into larger, organized collectives. Although all animals communicate, their channels are typically narrow and do not support expression of any and all thoughts. Language enables humans to move arbitrary thoughts from one mind to another, creating a new, cultural level of group organization. For most of human evolution, this system was local, allowing small bands of people to form local clusters of organization. Spoken language allowed hunter-gatherers to organize their foraging efforts and small farming communities their harvest, but not much more.

The origin of writing allowed the first large-scale societies, organized on hierarchical (often despotic) lines: A few powerful kings and scribes had control over the communication channels and issued edicts to all. This one-to-many topology is essentially endocrine. Despite their technological sophistication, radio and television share this topology. The proclamations and legal decisions of the ruler (or television producer) parallel the reproductive edicts carried by hormones within our bodies: commands issued to all, which all must obey.

Since Gutenberg, human society has slowly groped its way toward a new organizational principle. Literacy, mail, the telegraph, and democracy were steps along the way to a new organizational metaphor, more like the nervous system than like the hormones. The Internet completes the process: Now arbitrarily far-flung individuals can link, share information, and base their decisions on this new shared source of meaning. Like individual neurons in our neocortex, each human can potentially influence and be influenced, rapidly, by information from anyone, anywhere. We, the metaphoric neurons of the global brain, are on the brink of a wholly new system of societal organization, one spanning the globe with the metaphoric axons of the Internet linking us together.

The protocols are already essentially in place. TCP/IP and HTML are the global brain equivalents of cAMP and neurotransmitters:

universal protocols for information transfer. Soon a few dominant languages—say, English, Chinese, and Spanish—will provide for universal information exchange. Well-connected collective entities such as Google and Wikipedia will play the role of brain stem nuclei to which all other information nexuses must adapt.

Two main problems mar this "global brain" metaphor. First, the current global brain is only tenuously linked to the organs of international power. Political, economic, and military power remain insulated from the global brain, and powerful individuals can be expected to cling tightly to the endocrine model of control and information exchange. Second, our nervous systems evolved over 400 million years of natural selection, during which billions of competing false starts and miswired individuals were ruthlessly weeded out. But there is only one global brain today, and no trial-and-error process to extract a functional configuration from the trillions of possible configurations. This formidable design task is left up to us.

Search and Emergence

Rudy Rucker

Mathematician; computer scientist; cyberpunk pioneer; novelist; author, The Lifebox, the Seashell, and the Soul: What Gnarly Computation Taught Me About Ultimate Reality, the Meaning of Life, and How to Be Happy

Twenty or thirty years ago, people dreamed of a global mind that knew everything and could answer any question. In those early times, we imagined that we'd need a huge breakthrough in artificial intelligence to make the global mind work—we thought of it as resembling an extremely smart person. The conventional Hollywood image for the global mind's interface was a talking head on a wall-sized screen.

And now, in 2010, we have the global mind. Search engines, user-curated encyclopedias, images of everything under the sun, clever apps to carry out simple computations—it's all happening. But old-school artificial intelligence is barely involved at all.

As it happens, data and not algorithms are where it's at. Put enough information into the planetary information cloud, crank up a search engine, and you've got an all-knowing global mind. The answers emerge.

Initially people resisted understanding this simple fact. Perhaps this was because the task of posting a planet's worth of data seemed so intractable. There were hopes that some magically simple AI program might be able to extrapolate a full set of information from a few well-chosen basic facts—just as a person can figure out another person on the basis of a brief conversation.

At this point, it looks like there aren't going to be any incredibly concise *aha!*-type AI programs for emulating how we think. The good news is that this doesn't matter. Given enough data, a computer network can fake intelligence. And—radical notion—maybe that's what our wetware brains are doing, too. Faking it with search and emergence. Searching a huge database for patterns.

The seemingly insurmountable task of digitizing the world has been accomplished by ordinary people. This results from the happy miracle that the Internet is unmoderated and cheap to use. Practically anyone can post information on the Web, whether as comments, photos, or full-blown Web pages. We're like worker ants in a global colony, dragging little chunks of data this way and that. We do it for free; it's something we like to do.

Note that the Internet wouldn't work as a global mind if it were a completely flat and undistinguished sea of data. We need a way to locate the regions most desirable in terms of accuracy and elegance. An early, now discarded notion was that we would need some kind of information czar or committee to rank the data. But, here again, the anthill does the work for free.

By now it seems obvious that the only feasible way to rank the Internet's offerings is to track the online behaviors of individual users. By now it's hard to remember how radical and rickety such a dependence upon emergence used to seem. No control—what a crazy idea! But it works. No centralized system could ever keep pace.

An even more surprising success is found in user-curated encyclopedias. When I first heard of this notion, I was sure it wouldn't work. I assumed that trolls and zealots would infect all the posts. But the Internet has a more powerful protection system than I'd realized. Individual users are the primary defenders.

We might compare the Internet to a biological system in which new antibodies emerge to combat new pathogens. Malware is forever changing, but our defenses are forever evolving as well.

I am a novelist, and the task of creating a coherent and fresh novel always seems in some sense impossible. What I've learned over the course of my career is that I need to trust in emergence, also known as the muse. I assemble a notes document filled with speculations, overheard conversations, story ideas, and flashy phrases. Day after day, I comb through my material, integrating it into my mental Net, forging links and ranks. And, fairly reliably, the scenes and chapters of my novel emerge. It's how my creative process works.

In our highest mental tasks, any dream of an orderly process is a will-o'-the wisp. And there's no need to feel remorseful about this. Search and emergence are good enough for the global mind—and they're good enough for us.

My Fingers Have Become Part of My Brain

James O'Donnell

Classicist; provost, Georgetown University; author, The Ruin of the Roman Empire

How is the Internet changing the way I think? My fingers have become part of my brain. What will come of this? It's far too early to say.

Once upon a time, knowledge consisted of what you knew yourself and what you heard—literally, with your ears—from others. If you were asked a question in those days, you thought of what you had seen and heard and done yourself and what others had said to you. I'm rereading Thucydides this winter and watching the way everything depended on whom you knew, where the messengers came, from and whether they were delayed en route, walking from one end of Greece to another. Thucydides was literate, but his world hadn't absorbed that new technology yet.

With the invention of writing, the eyes took on a new role. Knowledge wasn't all in memory but was found in present, visual stimuli: the written word in one form or another. We have built a mighty culture based on all the things humankind can produce and the eye can study. What we could read in the traditional library of twenty-five years ago was orders of magnitude richer and more diverse than the most that any person could ever see, hear, or be told of in one lifetime. The modern correlative to Thucydides would be Churchill's history of World War II and the abundance of written documents he shows himself dependent on at every stage of the war. But imagine Churchill or Hitler with Internet-like access to information!

Now we change again. It's less than twenty years since the living presence of networked information has become part of our thinking machinery. What it will mean to us that vastly more people have nearly instantaneous access to vastly greater quantities of information cannot be said with confidence. In principle, it means a democratization of innovation and of debate. In practice, it also means a world in which many have already proved that they can ignore what they do not wish to think about, select what they wish to quote, and produce a public discourse demonstrably poorer than what we might have known in the past.

But just for myself, just for now, it's my fingers I notice. Ask me a good question today, and I find that I begin fiddling. If I am away from my desk, I pull out my BlackBerry so quickly and instinctively that you probably think I'm ignoring your question and starting to read my e-mail or play Brickbreaker—and sometimes I am. But when I'm not—that is, when you've asked a really interesting question—it's a physical reaction, a gut feeling that I need to start manipulating (the Latin root for "hand," *manus*, is in that word) the information at my fingertips to find the data that will support a good answer. At my desktop, it's the same pattern: The sign of thinking is that I reach for the mouse and start "shaking it loose"—the circular pattern on the mouse pad that lets me see where the mouse arrow is, make sure the right browser is open, get a search window handy. My eyes and hands have already learned to work together in new ways with my brain—in a process of clicking, typing a couple of words, clicking, scanning, clicking again—which really is a new way of thinking for me.

That finger work is unconscious. It just starts to happen. But it's how I can now tell thinking has begun, as I begin working my way through an information world more tactile than ever before. Will we next have three-dimensional virtual spaces in which I gesture, touch, and run my fingers over the data? I don't know; nobody can know.

But we're off on a new and great adventure whose costs and benefits we will only slowly come to appreciate.

What all this means is that we are in a different space now, one that is largely unfamiliar to us even when we think we're using familiar tools (like a "newspaper" that has never been printed or an "encyclopedia" vastly larger than any shelf of buckram volumes), and one that has begun life by going through rapid changes that only hint at what is to come. I'm not going to prophesy where that goes, but I'll sit here a while longer, watching the ways I've come to "let my fingers do the walking," wondering where they will lead.

A Mirror for the World's Foibles

John Markoff

Journalist; covers Silicon Valley for the New York Times; *author*, What the Dormouse Said: How the Sixties Counterculture Shaped the Personal Computer Industry

It's been three decades since Les Earnest, then assistant director of the Stanford Artificial Intelligence Laboratory, introduced me to the ARPANET. It was 1979, and from his home in the hills overlooking Silicon Valley he was connected via a terminal and a 2,400-baud modem to Human Nets, a lively virtual community that explored the impact of technology on society.

It opened a window for me into an unruly cyberworld that at first seemed to be, to quote computer music researcher and composer John Chowning, a Socratean abode. Over the next decade and a half, I joined the camp of what I have since come to think of as Internet utopians. The Net seemed to offer this shining city on a hill, free from the grit and foulness of the meat world. Ideologically, this was a torch carried by *Wired* magazine, and the ideal probably reached its zenith in John Perry Barlow's 1996 essay, "A Declaration of the Independence of Cyberspace."

Silly me. I should have known better. It would all be spelled out clearly in John Brunner's *The Shockwave Rider*, William Gibson's *Neuromancer*, Neal Stephenson's *Snowcrash*, Vernor Vinge's *True Names*, and even less-well-read classics such as John Barnes's *The Mother of Storms*. Science fiction writers were always the best social scientists, and in describing the dystopian nature of the Net they were again right on target.

There would be nothing even vaguely utopian about the reality of the Internet, despite preachy "The Road Ahead" vision statements by (late to the Web) luminaries like Bill Gates. This gradually dawned on me during the 1990s, driven home with particular force by the Kevin Mitnick affair. By putting every human on the planet directly in contact with every other, the Net opened a Pandora's box of nastiness.

Indeed, while it was true that the Net skipped lightly across national boundaries, the demise of localism didn't automatically herald the arrival of a superior cyberworld. It simply accentuated and accelerated both the good and the bad, in effect becoming a mirror for all the world's fantasies and foibles.

Welcome to a bleak *Blade Runner*–esque world dominated by Russian, Ukrainian, Nigerian, and American cybermobsters, in which our every motion and movement is surveilled by a chorus of Big and Little Brothers.

Not only have I been transformed into an Internet pessimist, but recently the Net has begun to feel downright spooky. Not to be anthropomorphic, but doesn't the Net seem to have a mind of its own? We've moved deeply into a world where it is leaching value from virtually every traditional institution in the name of some Borg-like future. Will we all be assimilated, or have we been already? Wait! Stop me! That was *The Matrix*, wasn't it?

a completely new form of sense

Terence Koh

Artist

i am very interested in the Internet, especially right now.

the Internet is a completely new form of sense.

as a human, i have experienced reality, as have the rest of my species since we had the ability to self-realize, as a combination of what we see, smell, feel, hear, and taste.

but the Internet—and this is a term i think that is beyond the idea of just the Web on a computer (Websites, e-mails, blogs, Twitter, Google, etc.) that is become "something" that i cannot myself really define yet—the Internet is really growing beyond this "something," so that even if someone does not have a computer, the Internet still affects them.

so this is very interesting, because the Internet is becoming a new form of sense that has not existed since we began to self-realize as humans.

and because this affects everybody, i feel that thinking about what the Internet is now must always come back to myself as an individual. because it is becoming more and more important to see how our individual thoughts and actions affect everything else around us. it all still starts with the "i," with me.

a new collective sense of "i" is the Internet . . .

so that there is a new form of "i" that is also "we" at the same time, because we are all involved with it.

i am not sure if i am answering your question, as it is a question that i do think about consciously every day now but can't quite figure out.

and forgive me if i sound like a bad science fiction writer, but if i may give any direction to your question, i think that the Internet is probably going to evolve by itself very very soon to give you better answers than i can ever give.

and i do not think i would even know it myself when that happens.

that is quite a scary thought.

By Changing My Behavior

Seirian Sumner

Research fellow in evolutionary biology, Institute of Zoology, Zoological Society of London

I was rather stumped by the *Edge* question, because I have little experience of work or play without the Internet. My interests and the way I think, work, and play have evolved alongside it. I thought it might help if I could work out what life would be like for me without the Internet. But abstaining from it is not a feasible experiment, even on a personal level. Instead, I exploited the very resource we are evaluating and asked my friends on Facebook what they thought their lives would be like without the Internet. If I could empathize with my alter ego in a parallel offline universe where there was no Internet, perhaps I could understand how the Internet has influenced the way I think.

Initial impressions of an Internet-free life from my Facebook friends were of general horror. The Internet plays a crucial role in our personal lives: My friends said they would be "lost," "stressed," "anxious," and "isolated" without it. They were concerned about:

"No 24/7 chats?"

"How would I make new friends/meet new people?"

"How would I keep in touch with my friends abroad?"

"I'd actually have to buy things in person from real people!"

We depend on the Internet as our social network, to connect with friends, or strangers, and to access resources. Sitting at my computer, I am one of the millions of nodes making up the network. Whereas physical interactions with other nodes in the network are largely impossible, I am potentially connected to them all.

Caution and suspicion of the unfamiliar are ancestral traits of humans, ensuring survival by protecting against usurpation and theft of resources. A peculiar thing about the Internet is that it makes us highly receptive and indiscriminate in our interactions with complete strangers. The other day I received a message inviting me to join a Facebook group for people sharing "Seirian" as their first name. Can I resist? Of course not! I'll probably never meet the seventeen other Seirians, but I am now a node connected to a virtual network of Seirians. Why did I join? Because I had nothing to lose, there were no real consequences, and I was curious to tap into a group of people wholly unconnected with my current social network. The more friendly connections I engage in, the greater the rewards I can potentially reap. If the Facebook Seirians had knocked on my real front door instead of my virtual one, would I have signed up? No, of course not—too invasive, personal, and potentially costly (they'd know where I live, and I can't unplug them). Contrary to our ancestral behaviors, we tolerate invasion of privacy online, and the success of the Internet relies on this.

Connectivity comes at the cost of privacy, but it does promote information acquisition and transfer. Although the initial response from my Facebook friends was fear of disconnection, more considered responses appreciated the Internet for the enormous resource it is, noting that it could never be replaced with traditional modes of information storage and transfer:

"How do I find things out?"

"Impossible to access information."

"You mean I have to physically go shopping/visit the library?"

"So slow."

"Small life."

The Internet relies on our craving for knowledge and connections but also on our astonishing online generosity. We show inordinate levels of altruism on the Internet, wasting hours on chat room sites giving advice to complete strangers or contributing anonymously to Wikipedia just to enrich other people's knowledge. There is no guarantee or expectation of reciprocation. Making friends and trusting strangers with personal information (be it your bank details or musical tastes) is an essential personality trait of an Internet user, despite being at odds with our ancestral natural caution. The data we happily give away on Facebook are exactly the sort of information that totalitarian secret police seek through interrogation. By relaxing our suspicions (or perception) of strangers and behaving altruistically (indiscriminately), we share our own resources and gain access to a whole lot more.

I thought I had too little pre-Internet experience to be able to answer this question. But now I realize that we undergo rapid evolution into a different organism every time we log on. The Internet may not necessarily change the way we think, but it certainly shapes and directs our thoughts by changing our behavior. Offline, we may be secretive, miserly, private, suspicious, and self-centered. Online, we become philanthropic, generous, approachable, friendly, and dangerously unwary of strangers. Online behavior would be selected out in an offline world, because no one would cooperate; people don't want unprompted friendship and generosity from complete strangers. Likewise, offline behavior does badly in an online world: Unless you give a little of yourself, you get restricted access to resources. The reason for our personality change is that the Internet is a portal to lazy escapism: At the twitch of the mouse, we enter a world in which the consequences of our actions don't seem real. The degree to which our online and offline personas differ will, of course, vary from one person to another. At the most extreme, online life is one of carefree fantasy: live vicariously through your flawless avatar in the fantastical world of Second Life. What better way to escape the tedium and struggles of reality that confront our offline selves?

Is the change from offline to online behavior adaptive? We ultimately strive to maximize our individual inclusive fitness. We can do this using our communication skills (verbal and written) to persuade other people to alter their behavior for mutual benefits. Early hominid verbal communication and hieroglyphs were the tools of persuasion used by our ancestors. The Internet is the third great breakthrough in human communication, and our behavioral plasticity is a necessary means for exploiting it. Do we need to moderate these shifts in behavior? One of my Facebook friends said that life would be "relaxing" without the Internet. Is our addiction to the Internet leaving us no time or space to think and process the complex stream of interactions and knowledge we get from it? Sleep is essential for "brain sorting"—maybe offline life (behavior) is, too.

To conclude: The Internet changes my behavior every time I log on and in so doing influences how I think. My daring, cheeky, spontaneous, and interactive online persona encourages me to think further outside my offline box. I think in tandem with the Internet, using its knowledge to inspire and challenge my thoughts. My essay is a testament to this: Facebook inspired my thoughts and provoked this essay, so I couldn't have produced it without the Internet.

There Is No New Self

Nicholas A. Christakis

Physician and social scientist, Harvard University; coauthor (with James H. Fowler), Connected: The Surprising Power of Our Social Networks and How They Shape Our Lives

Efforts to change the way we think—and to enhance our cognitive capacity—are ancient. Brain enhancers come in several varieties. They can be either hardware or software, and they can be either internal or external to our bodies. External hardware includes things such as cave paintings, written documents, eyeglasses, wristwatches, wearable computers, and brain-controlled machines. Internal hardware includes things such as mind-altering substances, cochlear implants, and intracranial electrical stimulation. Internal software includes things such as education, meditation, mnemonics, and cognitive therapy. And external software includes things such as calendars, voting systems, search engines, and the Internet.

I've had personal experience with most of these (save cave painting and the more esoteric forms of hardware), and I think I can say with confidence that they have not changed my brain.

What especially attracts my attention, though, is that the more complex types of external software—including the Internet—tend to involve communication and interaction, and thus they tend to be specifically social: They tend to involve the thoughts, feelings, and actions of many individuals, pooled in some way to make them accessible to individuals, including me. The Internet thus facilitates an age-old predilection of the human mind to benefit from our tendency as a species to be *Homo dictyous* (network man)—an innate tendency to connect with others and be influenced by them. In this regard, the Internet is both mind-expanding and atavistic.

The Internet is no different from previous, equally monumental brain-enhancing technologies, such as books or telephony, and I doubt whether books and telephony have changed the way I think, in the sense of actually changing the way my brain works (which is the particular way I am taking the question before us). In fact, it is probably more correct to say that our thinking gave rise to the Internet than that the Internet gave rise to our thinking. Another apt analogy may be mathematics. It has taken centuries for humans to accumulate mathematical knowledge, and I learned geometry and calculus in high school in a way that probably would have astonished mathematicians just a few centuries ago. But, like other students, I did this with the same brain we've all had for millennia. The math surely changed how I think about the world. But did it change the way I think? Did it change my brain? The answer is mostly no.

To be clear, the Internet is assuredly changing quite a few things related to cognition and social interaction. One widely appreciated and important example of both is the way the Internet facilitates hive-mind phenomena, like Wikipedia, that integrate the altruistic impulses and the knowledge of thousands of far-flung individuals. To the extent that I participate in such things (and I do), my thinking and I are both affected by the Internet.

But most thinking serves social ends. A strong indicator of this fact is that the intellectual content of most conversation is trivial, and it certainly is not focused on complex ideas about philosophy or mathematics. In fact, how often—unless we are ten-year-old boys—do we even think or talk about predators or navigation, which have ostensibly been important topics of thought and conversation for quite some time? Mostly we think and talk about each other. This is probably even true for those of us who spend our lives as scientists.

Indeed, our brains likely evolved their capacity for intelligence in response to the demands of social rather than environmental complexity. The evolution of larger social groups among primates required and benefited from the evolution of a larger neocortex (the outer, thinking part of our brain), and managing social complexity in turn required and benefited from the evolution of language. Known as the social-brain hypothesis, this idea posits that the reason we think at all has to do with our embeddedness in social life.

What role might technology play in this? Very little, it turns out. Consider, for example, the fact that the size of military units has not changed materially in thousands of years, even though our communication technology has, from signal fires to telegraphy to radio to radar. The basic unit in the Roman army (the maniple) was composed of 120 to 130 men, and the size of the analogous unit in modern armies (the company) is still about the same.

The fact that effective human group size has not changed very substantially—even though communication technology has—suggests that it is not the technology that is crucial to our performance. Rather, the crucial factor is the ability of the human mind to track social relationships, to form mental rosters that identify who is who, and to develop mental maps that track who is connected to whom and how strong or weak, or cooperative or adversarial, those relationships are. I do not think the Internet has changed the ability of my brain to do this. While we may use the word *friends* to refer to our contacts online, they are decidedly not our friends in the truly social, emotional, or biological sense of the word.

There is no new self. There are no new others. And so there is no new brain and no new way of thinking. We are the same species after the Internet as before. Yes, the Internet can make it easy for us to learn how to make a bomb or find a willing sexual partner. But the Internet itself is not changing the fundamental reality of my thinking any more than it is changing our fundamental proclivity to violence or our innate capacity for love.

I Once Was Lost but Now Am Found, or How to Navigate in the Chartroom of Memory

Neri Oxman

Architect, researcher, MIT; founder, Materialecology

> "I, myself, alone, have more memories than all mankind since the world began," he said to me. And also: "My dreams are like other people's waking hours." And again, toward dawn: "My memory, sir, is like a garbage heap."
>
> —"Funes el memorioso," Jorge Luis Borges

"Funes, His Memory" tells the evocative tale of Ireneo Funes, a Uruguayan boy who suffers an accident that leaves him immobilized along with an acute form of hypermnesia, a mental abnormality expressed in exceptionally precise memory. So vivid is Funes's memory that he can effortlessly recall the exact appearance of any physical object at each time he saw it. In his perpetual present, images unfold their archaeology as infinite wells of detailed information: "He knew the forms of the clouds in the southern sky on the morning of April 30th, 1882." Funes's memories are intensely present, as muscular and thermal sensations accompanying every visual record. He can reconstruct every event he has ever experienced. His recollections are so accurate that the time it takes to reconstruct an entire day's worth of events equals the duration of the day itself. In Funes's world, reflection makes no sense at all, as there is simply no time or motive to reflect or interpret.

As a consequence, Funes is unable to suppress details, and any attempt to conceive of, or manage, his impressions—the very stuff of thought—is overridden by relentlessly literal recollections ("We, in a glance, perceive three wine glasses on the table; Funes saw all the shoots, clusters, and grapes of the vine"). Funes is not able to generalize or to deduce or induce anything he experiences. Things are just what they are, scaled one-to-one. Cursed with meticulous memory, he escapes to remoteness and isolation—a "dark room"—where new images do not enter and where, motionless, he is absorbed in the contemplation of a sprig of artemisia.

Hypermnesia appears to have been to Funes what the World Wide Web is today to the human race. An inexhaustible anthology of every possible thing recorded at every conceivable location in any given time, the Internet is displacing the role of memory, and it does so immaculately. Any imaginable detail about the many dimensions of any given experience is either recorded or consumed as yet another fragment of reality. There is no time to think, it seems. Or perhaps this is just a new kind of thinking. Is the Web yet another model of reality, or is reality becoming a model of the Web?

In his "On Exactitude in Science," Borges carries on with similar ideas, describing an empire in which the craft of cartography attained such precision that its map has emerged as large as the kingdom it depicts. Scale, or difference, has been replaced by repetition. Such a map embodies the dissimilarity between reality and its representation. It becomes the territory itself, and the original loses authenticity. It achieves the state of being more real than real, as there is no reality left to chart.

The Internet, no doubt, has become such a map of the world, both literally and symbolically, as it traces in almost 1:1 ratio every event that has ever taken place. One cannot afford to get lost in a space so perfectly detailed and predictable. Physical navigation is solved, as online maps offer even the most exuberant flaneur the knowledge of prediction. But there are also enormous mental implications to this.

As we are fed with the information required, or desired, to understand and perceive the world around us, the power of perception withers, and the ability to engage in abstract and critical thought atrophies. Models become the very reality we are asked to model.

If one believes that the wetware source of intellectual production, whether in the arts or sciences, is guided by the ability to critically model reality, to scale information, and to engage in abstract thought, where are we heading in the age of the Internet? Are we being victimized by our own inventions? The Internet may well be considered an oracle, the builder of composite and hybrid knowledge—but is its present instantiation inhibiting the cognitive nature of reflective and creative thought?

Funes is portrayed as an autistic savant, with the gift of memorizing anything and everything. This gift eventually drives him mad, but Borges is said to have constructed Funes's image to suggest the "waste of miracle" and point at the vast and dormant potential we still encompass as humans. In letting the Internet think for us, as it were, are we encouraging the degeneration of our mental capacities? Is the Internet making us obliviously somnolent?

Between the associative nature of memory and the referential eminence of the map lies a blueprint for the brain. In the ambience of future ubiquitous technologies looms the promise of an ecstasy of connectivity (or thus is the vision of new consciousness à la Gibson and Sterling). If such a view of augmented interactivity is even remotely accurate (as it must be), it is the absence of a cognate presence that defies the achievement of transforming the Internet to a new reality, a universally accessible medium for enhanced thinking. If the Internet can become an alternative medium of human consciousness, how can a cognate presence inspire the properties of infinite memory with the experiential and the reflective, all packaged for convenience and pleasure in a Mickey Mouse–like antenna cap?

In Borges's tale, Funes cites a revealing line from Pliny's *Naturalis Historia* in the section on memory. It reads, "Ut nihil non iisdem verbis redderetur auditum": "So that nothing that has been heard can be retold in the same words."

The Greatest Pornographer

Alun Anderson

Senior consultant and former editor in chief and publishing director of New Scientist; *author*, After the Ice: Life, Death, and Geopolitics in the New Arctic

The Internet may not have changed how my brain works, but if you take *thinking* to mean the interaction between what's in your brain, what's in other people's brains, and what's in the environment around you, then the Internet is changing everything. In my line of work, as a writer and journalist, "changing the way you think" is now more of an imperative than a possibility: If you don't change, you risk extinction.

Powerful new technologies inevitably work a destructive fire on older ways. As advertising revenues vanish to the Internet, newspapers and magazines find they can no longer subsidize the information-gathering operations that the public is unwilling to pay for directly. The job of print journalist is starting to look as quaint as that of chimney sweep. Many of the print newspapers and magazines that employ those journalists may not survive the Internet at all.

The book is likely set to vanish, too. I imagine a late-twenty-first-century Wikipedia entry reading:

> BOOK: A format for conveying information consisting of a single continuous piece of text, written on an isolated theme or telling a particular story, averaging around 100,000 words in length and authored by a single individual. Books were printed on paper between the mid-fifteenth and early twenty-first centuries but more often delivered electronically after 2012. The book largely disappeared during the mid-twenty-first century, as it became clear that it had never been more than a narrow in-

stantiation, constrained by print technology, of texts and graphics of any form that could flow endlessly into others. Once free from the shackles of print technology, new storytelling modes flowered in an extraordinary burst of creativity in the early twenty-first century. Even before that, the use of books to explain particular subjects (see textbook) had died very rapidly, as it grew obvious that a single, isolated voice lacked authority, wisdom, and breadth.

These changes and wonderful new creative opportunities, arrived or arriving, are the outward manifestation of a change in how we think as we shift away from information scarcity, low levels of interpersonal interaction, and little feedback on the significance of what we say, to information abundance and high levels of interaction and feedback.

As a journalist, I can remember when my most important possession was a notebook of "contacts." The information I wrested from them was refined with the help of a few close colleagues. That is the past. Thanks to the Internet, search engines, and the millions of organizations, pressure groups, and individuals who are producing free information, almost everything is already out there and available to everyone.

My work is not digging out information but providing the narrative thread that connects it. In the deluge of bits, it is the search for the bigger picture, the larger point, that matters. You no longer find things out but find out what they mean. That new way of thinking is not so easy. Even the mighty U.S. Department of Homeland Security could not connect the dots regarding a recent incident when different fragments of information about a young Nigerian radical surfaced. As a result, a plane full of passengers was very nearly blown from the sky.

To do that job well, I don't think with just a few close colleagues; I've delocalized my thought and spread it around the world electronically. (Homeland Security might need to do the same.) With the Internet, my thoughts develop through sharing them with others who have a like interest; I have virtual friendships of ideas with scores of people I probably will never meet and whose age, background, and gender I do not even know. Their generosity is a delight. Anything I write is now soon modified. I don't think alone. Rather, I steer a global conversation given form by the Web.

Neither magazines nor books in solid, physical form are good at capturing this flow, which is partly why their future is uncertain. The survivors among them may be those that exult in their physicality—in their existence as true objects. Physical beauty will flourish alongside a virtual world. I look forward to a rebirth for magazines with a touch, feel, look, and smell that will make them a pleasure to hold closely.

The word *pleasure* is a good one with which to switch direction. The Internet may be changing the way I think in the cerebral sense, but it may be changing the way the world thinks in a far more physical way. The Internet is awash with sex. In a few hours, an innocent can see more of the pleasures and perversions of sex, in a greater number of close-up couplings, than an eighteenth-century roué could experience in a lifetime devoted to illicit encounters. The Internet is the greatest sex education machine—or the greatest pornographer—that has ever existed. Having spent time teaching at a Muslim university, where the torrent of Internet sex was a hot topic, I would not underestimate its impact on traditional societies. There is a saying that rock and roll brought down the Soviet Union; once the Soviet subconscious had been colonized, the political collapse followed easily. The flood of utterly uncensored images of sexual pleasure that reaches every corner of the world is certainly shaking the thinking of young men and women in the conservative societies I've worked in. Where the conflicting emotions that have been unleashed will lead, I cannot tell.

My Sixth Sense

Albert-László Barabási

Distinguished Professor and director of Northeastern University's Center for Complex Network Research; author, Bursts: The Hidden Pattern Behind Everything We Do

For me the Internet is more than a search engine: It has become the subject of my research, a proxy of the many complex systems we are surrounded with.

I know when this transition started. It was December 1994—which was the time I decided to learn a bit about computers, given that my employer at that time was IBM. So I lifted a book about computer science from a shelf in the Thomas J. Watson Research Center to keep my mind engaged during the holidays. It was my first encounter with networks. A few months later, I submitted my first research paper on the subject, and it was promptly rejected by four journals. No one said it was wrong. The common response was: "Why should we care about networks?" (While my paper never got published, it is still available—where else but on the Internet? At the Los Alamos preprint archive, to be precise.)

The Internet eventually rescued me, but it took four more years. In the meantime, I sent countless e-mails to search engines asking for data on the topology of the Web. (All those requests must still be on their way to V4641 Sgr, the closest black hole to Earth, somewhere out there in the Milky Way.) Finally, in 1998 a gifted postdoc, Hawoong Jeong, told me that he knew how to build a search engine. And he did, providing us the map that has finally legitimized my years of persistence and serial failures. In 1999 it led to my first publication on networks. It was about the structure of the World Wide Web.

Today my work could not be possible without the Internet. I do not mean only the access to information. The Internet has fundamentally changed the way I approach a research problem. Much of my research consists of finding organizing principles—laws and mechanisms—that apply not to one but to many complex systems. If these laws are indeed generic and universal, they should apply to our online world as well, from the Internet to online communities on the World Wide Web. Thus, we often test our ideas on the Internet, rather than in the cell or in economic systems, which are harder to monitor and measure.

The Internet is my sixth sense, altering the way I approach a problem. But it has just as fundamentally changed what I think about, and that may be even more significant in the end.

The Internet Reifies a Logic Already There

Tom McCarthy

Artist and novelist; author, C

"How has the Internet changed the way you think?" It hasn't.

Western culture has always been about networks. Look at Clytemnestra's "beacon telegraph" speech in the *Oresteia*, or the relay system of oracles and cryptic signals that Oedipus has to navigate. Look at Daniel Schreber's vision of wires and nerves, or Kafka's and Rilke's visions of giant switchboards linking mortals to (and simultaneously denying them access to the source code of) gods and angels. Or the writings of Heidegger or Derrida: meshes, relays, endless transmission. The Internet reifies a logic that was already there.

Instant Gratification

Peter H. Diamandis

Chairman/CEO, X PRIZE Foundation

In mid-2009, I made a seven-day, round-the-world business trip from Los Angeles to Singapore, India, United Arab Emirates, and Spain. It was a lecture tour—all work. As I landed in each of these countries, I tried an experiment and tweeted my landing, asking if any friends were in that country. My tweet was automatically posted to my Facebook. In each case, in each country, my inquiry was answered with a "Hey, I happen to be in town as well. Let's meet for coffee." Instant and very unexpected gratification. Ask and you shall receive.

In a separate experiment, I was musing about the volume of gold mined by human beings since the start of mining industry. I was interested because I was fascinated with the idea of mining precious metals from asteroids in the decades ahead. I had done some back-of-the-envelope calculations that amazed me. I posted the following: "Total gold ever mined on Earth is 161,000 tons. Equal to ~20 meters cubed . . . pls check my math!!"

Within minutes, I had three confirmations of the calculation, as well as numbers for platinum (~6 meters cubed), rhodium (~3 meters cubed) and palladium (~7 meters cubed). Ask and you shall receive.

How many times do I wonder about something and then let it drop? I'm realizing that even complex questions can be answered (with enhancements!) with little more work than a digital prayer cast into the socialverse. The better and more intriguing my questions, the more compelling the answers I receive. Looking forward, I can imagine this holding true for requests for artwork, videos, manufactured goods. The point is that nearly instantaneous gratification is possible, and it's the quality of the incentive that is most important, incentives in this case being a chance encounter and an intriguing question. Future incentives will be such things as cash, or who is asking the question, or the importance of the problem to be solved.

The Internet as Social Amplifier

David G. Myers

Social psychologist, Hope College; author, A Quiet World: Living with Hearing Loss

I cut my eyeteeth in social psychology with experiments on group polarization—the tendency for face-to-face discussion to amplify group members' preexisting opinions. Never then did I imagine the potential dangers, or the creative possibilities, of polarization in virtual groups.

Electronic communication and social networking enable Tea Partiers, global warming deniers, and conspiracy theorists to isolate themselves and find support for their shared ideas and suspicions. As the Internet connects the like-minded and pools their ideas, white supremacists may become more racist, Obama despisers more hostile, and militia members more terror-prone (thus limiting our power to halt terrorism by conquering a place). In the echo chambers of virtual worlds, as in real worlds, separation + conversation = polarization.

But the Internet as social amplifier can instead work for good, by connecting those coping with challenges. Peacemakers, cancer survivors, and bereaved parents find strength and solace from kindred spirits.

By amplifying shared concerns and ideas, Internet-enhanced communication can also foster social entrepreneurship. An example: As a person with hearing loss, I advocate a simple technology that doubles the functionality of hearing aids, transforming them, with the button push, into wireless loudspeakers. After experiencing this "hearing loop" technology in countless British venues, from cathedrals to post office windows and taxis, I helped introduce it to western Michigan, where it can now be found in several hundred venues, including Grand Rapids' convention center and all gate areas of its airport. Then, via a Website, hearing listservs, and e-mail, I networked with fellow hearing advocates, and by feeding one another our resolve gained strength.

Thanks to the collective efficacy of our virtual community, hearing-aid-compatible assistive listening has spread to other communities and states. New York City is installing it in 488 subway information booths. Leaders in the American Academy of Audiology and the Hearing Loss Association of America are discussing how to promote this inexpensive, wireless assistive listening. Several state hearing-loss associations are recommending it. The hearing industry is now including the needed magnetic receiver in most hearing aids and cochlear implants. And new companies have begun manufacturing and marketing hearing-loop systems. Voilà—a grassroots, Internet-fueled transformation in how America provides listening assistance is under way.

The moral: By linking and magnifying the inclinations of kindred-spirited people, the Internet can be very, very bad, but also very, very good.

Navigating Physical and Virtual Lives

Linda Stone

High-tech industry consultant; former executive, Apple Computer and Microsoft Corporation

Before the Internet, I made more trips to the library and more phone calls. I read more books and my point of view was narrower and less informed. I walked more, biked more, hiked more, and played more. I made love more often.

The seductive online sages, scholars, and muses that joyfully take my curious mind wherever it needs to go, wherever it can imagine going, whenever it wants, are beguiling. All my beloved screens offer infinite, charming, playful, powerful, informative, social windows into global human experience.

The Internet, the online virtual universe, is my jungle gym, and I swing from bar to bar, learning about how writing can be either isolating or social, about DIY Drones (unmanned aerial vehicles) at a Maker Faire, about where to find a quantified-self meetup, about how to make *sach moan sngo num pachok*. I can use an image search to look up "hope" or "success" or "play." I can find a video on virtually anything: I learned how to safely open a young Thai coconut from this Internet of wonder.

As I stare out my window at the unusually beautiful Seattle weather, I realize I haven't been out to walk yet today—sweet Internet juices still dripping down my chin. I'll mind the clock now, so I can emerge back into the physical world.

The physical world is where I not only see but also feel—a friend's loving gaze in conversation, the movement of my arms and legs and the breeze on my face as I walk outside, and the company of friends for a game night and potluck dinner. The Internet supports my thinking, and the physical world supports that as well as rich sensing and feeling experiences.

It's no accident that we're a culture increasingly obsessed with the Food Network and farmers' markets—they engage our senses and bring us together with others.

How has the Internet changed my thinking? The more I've loved and known it, the clearer the contrast, the more intense the tension between a physical life and a virtual life. The Internet stole my body, now a lifeless form hunched in front of a glowing screen. My senses dulled as my greedy mind became one with the global brain we call the Internet.

I am confident that I can find out about nearly anything online and also confident that in my time offline I can be more fully alive. The only tool I've found for this balancing act is intention.

The sense of contrast between my online and offline lives has turned me back toward prizing the pleasures of the physical world. I now move with more resolve between each of these worlds, choosing one, then the other—surrendering neither.

Not Everything or Everyone in the World Has a Home on the Internet

Barry C. Smith

Professor and director, Institute of Philosophy, School of Advanced Study, University of London

The growth of the Internet has reversed previous assumptions: The private is now public; the local appears globally; information is entertainment; consumers turn into producers; everyone is an expert; the socially isolated become part of an enormous community preferring the virtual to the real. What have all these changes brought about?

Initially they appear empowering. Everyone can have a say, opinion is democratic, and at a time when natural resources are shrinking and environmental threats require us to limit our emissions, the Internet seems to be an ever expanding and almost limitless resource. Here, it seems, I discover a parallel world, where neat models replace messy reality, where freedom reigns, where wrongs are righted and fates can be changed. I am cheered by the possibilities.

However, the truth is that the virtual world grows out of, and ultimately depends on, the one world whose inputs it draws on, whose resources it consumes, and whose flaws it inevitably inherits. I find everything there: the good, the bland, the important, the trivial, the fascinating, and the off-putting. And just as there are crusading writers and eyewitness reporters, there are also cyber lynch mobs, hate-mailers, and stalkers. As more of my information appears on the Net, more use is made of it, for good or for ill. Increasing Internet identity means increasing identity theft, and whatever I have encrypted, hackers will try to decode. So much so that governments and other organizations often restrict their most secure communications to older technologies, even sending scrolled messages in small capsules through pneumatic tubes. This, of course, fuels the suspicions of Internet conspiracy theorists.

Looking at what have I've gained: I now hear from a greater range of different voices, discover new talents with something to say—niche writers, collectors, musicians, and artists. I have access to more books, journal articles, newspapers, TV programs, documentaries, and films. Missed something live? It will be on the Web. The greatest proportion of these individuals and outputs were already offering something interesting or important, to which the Internet gave worldwide access. Here we have ready-made content for the voracious Internet to consume and display.

But new media have emerged, too, whose content arose for, or on, the Internet. These include blogging, Wikipedia, and YouTube, along with new forms of shared communication, such as Facebook, Google Groups, and Twitter. Will these new forms replace the ready-made contents? It's unclear. Amid the bread-and-circuses element of the Internet, there is a need for good-quality materials and a means to sort the wheat from the chaff. Garbage in, garbage out, as computer programmers say. It is our choice, some will argue, yet I find myself looking with sheer disbelief, or ironic amusement, at what people have chosen to put up on the Net. The most fascinating are bloggers who provide alternative slices of life. Here we have diarists who desire to be intimate with everyone. Those with a distinctive voice and a good theme have found a following; when worldwide word spreads, the result is usually a contract to publish their output, lightly edited, as a book, which in turn can be read on the Internet.

What of the new Web-dependent phenomena: open access and open-source programming, virtual social networking, the co-construction of knowledge? All these are gains and reflect something hopeful: the collaborative effort of our joint endeavor, our willingness to share. The inclusive nature of these phenomena is encouraging. I want to join in, and I like the idea of making a modest contribution to a larger enterprise. But the new technologies let me witness their distancing and distorting influences: Internet-fueled fantasies where everyone can be a celebrity, or people can live through their avatars in virtual reality or develop alternative personalities in chat rooms—fantasies that someone, somewhere on the Internet, is making money from.

How do I cope with the speeded-up information age? The overload is overwhelming, but so is my desire to know and not to miss anything. I'm tempted to know a little bit about everything and look for predigested, concise, neatly formatted content from reliable sources. My reading habits have changed, making me aware of how important well-packaged information has become. It's now necessary to consume thousands of abstracts from scientific journals, doing one's own fast search for what should be read in more detail. Debates seem to be decided at the level of abstracts, repudiations signaled by the title and a hundred words. The real work, of course, goes on elsewhere, but we want the Internet to brings us the results. This leaves me knowing less and less about more and more. At the same time, I'm exhilarated by the dizzying effort to make connections and integrate information. Learning is faster—though the tendency to forge connecting themes can feel dangerously close to the search for patterns that overtakes the mentally ill. Time to slow down and engage in longer study.

The Internet shows me more and more about those who participate in it, but I worry lest I forget that not everything or everyone in the world has a home on the Internet. Missing are those who cannot read or write, who have no access to a computer, or who choose to remain disconnected. There is a danger of coming to think that what cannot be found on an Internet search doesn't exist, and that the virtual world is *the* world. It isn't. However bizarre and incredible the people populating the Internet are, they are still akin to me—people with knowledge of computers and their applications. Certainly there is diversity and hierarchy and vast domains of varied information, but nevertheless—except when Internet users turn their attention on the those who are excluded, or who exclude themselves—a mirror will be held up to those who sustain the information age, and it is only this part of the world that I have come to have scattered information about.

Ephemera and Back Again

Chris DiBona

Open Source and Public Sector, Google

I often feel as though my brain is at best a creative and emotional caching front end on the Internet. With a few bare exceptions (my children, my wife, my family), I feel little practical need anymore to commit my long-term memory to endeavors I formerly spent days, weeks, months, and years on. I've come to think I should memorize things more for the health of my brain rather than for any real practical need to know—for example, that decimal 32 is a space in ASCII, or that the second stanza of the Major General's song shows his acquaintance with the binomial theorem.

I hardly ever memorize phone numbers of people outside my immediate family these days, and I used to proudly tuck away nearly all of them. Now, as a result of the richness of a life connected to the Internet, I mostly retain area codes, so that I can guess who might be calling. A casualty of contact syncing, perhaps, but still I find myself considering many voice conversations or audio recordings too information-sparse to be listened to unless I'm otherwise occupied—with driving, say, or washing the dishes.

For elements of culture especially, I don't wonder for long who was in the movie about the fall of communism with the woman in a coma. I just look it up, faster, online. I don't spend much time considering in what techno song the dude from *Star Trek* says, "Time becomes a loop," nor do I find it difficult to find, online, the name of that book I read in which the dude orbiting a neutron star for an alien race discovers its tidal effects. Nor do I have to consider what game it was that had a dog accompanying me through post-apocalyptic California. As I scroll, Pavlovian, through my feed, the waves of knowledge roll over me.

When I travel, I no longer take pictures of these outings, except ones with my family in them; there are better photos available to me online, if I feel like jogging my memory about a trip.

I don't even especially worry about where I am, either, considering myself not unlike a packet being routed—not from client machine to router to server to backhaul to peer to machine to client machine, but instead from house to car to plane to car to hotel to car to office or conference to car to hotel to car to plane to car to home, with jet lag my only friend and my laptop my source of entertaining books (*Neutron Star*), movies (*Good Bye Lenin!*), games (Fallout), or music (Orbital, Meat to Munich), with cellular data, headphones, and circuits.

Some would equate this sort of information pruning to a kind of reinforced and embraced ignorance or evidence of an empty life. Nicholas Carr, writing in the *Atlantic*, enjoyed some attention in 2008 with his article titled "Is Google Making Us Stupid?" The author, reacting to (or justifying) his own reduced attention span, accuses Google (my employer) of trying to do away with deep thinking, while indulging in what comes off as an absurd nostalgia for making knowledge difficult to find and obtain.

There was an important thought worthy of exploration within that article—that there is a kind of danger in reinforcing the shallow. I have come to understand, expect, and accept that people try to find the Internet that aligns with their beliefs. This is impossible to change without strangling the Internet of the creativity that makes it so useful, as for every Wikipedia expanding and storing humankind's knowledge about everything, there is a Conservapedia rewriting the Bible to be more free-market friendly.

But people who wallow in ignorance are no different online than off. I don't believe that the Internet creates ignorant people. But what the Internet changes is the notion of unique thought. I have come to think that with more than 6.7 billion people on the planet, with more than a billion capable of expressing themselves on the Internet and hundreds of millions if not billions on the Internet via their cell phones, there is very little chance that any idea I might have outside my specialty hasn't already been explored, if not executed. Within my specialty, even, there is a fair amount of what I'd charitably call non-unique thinking. This is not to say that the world doesn't need practitioners. I consider myself to be a good one, but only rarely do I come up with an approach that I'd consider unique within my specialty.

At one time, I found this a rather bleak realization—thinking we're all just conduits from urge to hand to Net to work—but over the last decade I've come to find it a source of comfort. Not all ideas need be mine; I can use the higher functions more for where they matter—locally, with my family, and on my work, on things I enjoy and treasure—and less on loading a browser or opening a tab into today's ephemera.

Queries I executed while writing this article:

modern major general
Google stupid
garden paving pruning cleaving
garden paring pruning cleaving
garden paring pruning
garden paring
dense antonyms
major general's song
define: stanza
ASCII chart
game had a dog accompanying me through post-apocalyptic California
orbiting a neutron star for an alien race finds out about tidal effects
for an alien race finds out about tidal effects
orbiting a neutron star in a ship built by aliens

dude orbiting a neutron star for an alien race with eyes in their hands
time becomes a loop
the German movie about the fall of communism with the woman in a coma
books printed each year
Conservapedia
Internet enabled cell phones
people with Internet enabled cell phones
people with Internet-enabled cellphones
planet population

What Do We Think About? Who Gets to Do the Thinking?

Evgeny Morozov

Commentator on Internet and politics, Net Effect *blog; contributing editor,* Foreign Policy

As it might take decades for the Internet to rewire how our brains actually process information, we should expect that most immediate changes would be social rather than biological in nature. Of those, two bother me in particular. One has to do with how the Internet changes what we think about; the other one with who gets to do the thinking.

What I find particularly worrisome with regard to the "what" question is the rapid and inexorable disappearance of retrospection and reminiscence from our digital lives. One of the most significant but overlooked Internet developments of 2009—the arrival of the so-called real-time Web, whereby all new content is instantly indexed, read, and analyzed—is a potent reminder that our lives are increasingly lived in the present, completely detached even from the most recent past. For most brokers dealing on today's global information exchange, the past is a "strong sell."

In a sense, this is hardly surprising. The social beast that has taken over our digital lives has to be constantly fed with the most trivial of ephemera. And so we oblige, treating it to countless status updates and zetabytes of multimedia (almost 1,000 photos are uploaded to Facebook every second). This hunger for the present is deeply embedded in the very architecture and business models of social networking sites. Twitter and Facebook are not interested in what we were doing or thinking about five years ago; it's what we are doing or thinking about right now that they like to know.

These sites have good reasons for such a fundamentalist preference for the present, as it greatly enhances their ability to sell our online lives to advertisers. After all, much of the time we are thinking of little else but satisfying our needs, spiritual or physical, and the sooner our needs can be articulated and matched with our respective demographic group, the more likely it is that we'll be coerced into buying something online.

Our ability to look back and engage with the past is one unfortunate victim of such reification of thinking. Thus, amid all the recent hysteria about the demise of forgetting in the era of social networking, it's the demise of reminiscence that I find deeply troublesome. The digital age presents us with yet another paradox: Whereas we have nearly infinite space to store our memories, as well as all the multipurpose gadgets to augment them with GPS coordinates and 360-degree panoramas, we have fewer opportunities to look back on and engage with those memories.

The bottomless reservoirs of the present have blinded us to the positive and therapeutic aspects of the past. For most of us, "re-engaging with the past" today means nothing more than feeling embarrassed over something we did years ago after it has unexpectedly resurfaced on social networks. But there is much more to reminiscence than embarrassment. Studies show that there is an intricate connection between reminiscence (particularly about positive events in our lives) and happiness: the more we do of the former, the more we experience the latter. Substituting links to our Facebook profiles and Twitter updates for links to our past risks turning us into hyperactive, depressive, and easily irritated creatures who don't know how to appreciate our own achievements.

The "who" question—who gets to do the thinking in the digital age?—is much trickier. The most obvious answer—that the Internet has democratized access to knowledge, and we are all thinkers now, bowed over our keyboards much like Rodin's famous sculpture—is wrong. One of my greatest fears is that the Internet will widen the gap between the disengaged masses and the overengaged elites, thus thwarting our ability to collectively solve global problems (climate change and the need for smarter regulation in the financial industry come to mind) that require everyone's immediate attention. The Internet may yield more "thinking" about such issues, but such thinking will not be equally distributed.

The Marxists have been wrong on many issues, but they were probably right about the reactionary views espoused by the lumpen proletariat. Today we are facing the emergence of the cyber lumpen proletariat—of people who are being sucked into the digital whirlwind of gossip sites, trashy video games, populist and xenophobic blogs, and endless poking on social networking sites. The intellectual elites, on the other hand, continue thriving in the new digital environment, exploiting superb online tools for scientific research and collaboration, streaming art house films via Netflix, swapping their favorite books via e-readers, reconnecting with musical treasures of the bygone eras via iTunes, and, above all, perusing materials in the giant online libraries, like the one that Google could soon unveil. The real disparities between the two groups become painfully obvious once members of the cyber lumpen proletariat head to the polls and push for issues of an extremely dubious, if not outright unethical, nature. (The referendum on minarets in Switzerland is a case in point; the fact that Internet users on Obama's change.gov site voted the legalization of marijuana as the most burning issue is another.)

As an aside, given the growing concerns over copyright and the digitization of national cultural heritage in many parts of the world, there is a growing risk that this intellectual cornucopia will be available only in North America, creating yet another divide. Disconnected from Google's digital library, the most prestigious universities in Europe or Asia may look less appealing even than middling community colleges in the United States. It seems increasingly likely that the Internet will not diffuse knowledge production and thinking around the globe but further concentrate it in one place.

The Internet Is a Cultural Form

Virginia Heffernan

Columnist ("The Medium"), the New York Times

People who study the real world, including historians and scientists, may find that the reality of the Internet changes how they think. But those of us, including philosophers and literary critics, who study symbolic systems find in the Internet yet another symbolic system, albeit a humdinger, that yields—spectacularly, I must say—to our accustomed modes of inquiry.

Anyway, a new symbolic order need not disrupt Truth, wherever Truth may now be said to reside (neurons? climate change? atheism?). Certainly to those of us who read more novels than MRIs, the Internet—and especially the World Wide Web—looks like what we know: a fictional world made mostly of words.

Philosophers and critics must only be careful, as we are trained to be careful, not to mistake this highly stylized and artificial order, the Internet, for reality itself. After all, cultural vocabularies that gain currency and influence—epic poetry, the Catholic Mass, the British Empire, photography—*always* do so by purporting to be reality, to be transparent, to represent or circumscribe life as it really is. As an arrangement of interlocking high, pop, and folk art forms, the Internet is no different. This ought to be especially clear when what's meant by "the Internet" is that mostly comic, intensely commercial, bourgeois space known as the World Wide Web.

We who have determinedly kept our heads while suffrage, the Holocaust, the highway system, Renaissance perspective, coeducation, the Pill, household appliances, the moon landing, the Kennedy assassination, and rock and roll were supposed to change existence forever cannot falter now. Instead of theatrically changing our thinking, this time we must keep our heads, which means—to me—that we must keep on reading and not mistake new texts for new worlds, or new forms for new brains.

Wallowing in the World of Knowledge

Peter Schwartz

Futurist, business strategist; cofounder, Global Business Network, a Monitor Company; author, Inevitable Surprises

In 1973, just as I was starting work at Stanford Research Institute (SRI), I had the good fortune to be one of the earliest users of what was then known as ARPANET. Collaborative work at a distance was the goal of the experiment that led to the suitcase-size TI Silent 700 portable terminal with an acoustic coupler and thermal printer on the back (no screen) sitting on my desk at home in Palo Alto. I was writing scenarios for the future of Washington state with the staff of Governor Dan Evans in Olympia. It was the beginning of the redistribution of my sense of identity.

In the 1980s, I was also a participant in the WELL, one of the first meaningful online communities. Nearly everyone who was part of the WELL had this sense of a very rich set of multiple perceptions constantly and instantly accessible. And (though not because the Deadheads were a large part of that community) my sense of an aware, distributed consciousness began to develop.

Finally, with the coming of the modern Internet, the World Wide Web, and the incredible explosion of knowledge access, another level in transformation took hold. I am one of those people who used to read encyclopedias and almanacs. I just wanted to know more—actually, everything. I also make my living researching, writing, speaking, and consulting. Depth, breadth, and richness of knowledge are what make it work in my passions and my profession. Before the Internet, that was limited by the boundaries of my brain. Now there is a nearly infinite pool of accessible information that becomes my knowledge in a heartbeat measured in bits per second. For those of us who wallow in the world of knowledge for pleasure and profit, the Internet has become a vast extension of our potential selves.

The modern Internet has achieved much of what Ted Nelson articulated decades ago in his vision of the Xanadu project, or Doug Engelbart in his human augmentation vision at SRI. Nearly all useful knowledge is now accessible instantaneously from much of the world. Our effective personal memories are now vastly larger—essentially infinite. Our identity is embedded in what we know. And how I think is an expression of that identity.

For me, the Internet has led to that deep sense of collaboration, awareness, and ubiquitous knowledge that means that my thought processes are not bound by the meat machine that is my brain, nor my locality, nor my time.

One's Guild

Stewart Brand

Founder, Whole Earth Catalog; *cofounder,* the WELL; *cofounder,* Global Business Network; *author,* Whole Earth Discipline

I couldn't function without them, and I suspect the same is true for nearly all effective people. By "them," I mean my closest intellectual collaborators. They are the major players in my social, extended mind. How I think is shaped to a large degree by how they think.

Our association is looser than a team but closer than a cohort, and it's not a club or a workgroup or an elite. I'll call it a guild. The members of my guild run their own operations, and none of us reports to any other. All we do is keep close track of what the others are thinking and doing. Often we collaborate directly, but most of the time we don't. Everyone in my guild has his or her own guild, each of which is largely different from mine. I'm probably not considered a member of some of them. My guild nowadays consists of Danny Hillis, Brian Eno, Peter Schwartz, Kevin Kelly, John Brockman, Alexander Rose, and Ryan Phelan. Occasionally we intersect institutionally via the Long Now Foundation, Global Business Network, or Edge.org.

One's guild is a conversation extending over years and decades. I hearken to my gang because we have overlapping interests, and my gang keeps surprising me. Familiar as I am with them, I can't finish their sentences. Their constant creativity feeds my creativity, and I try to do the same for them. Often the way I ponder something is to channel my guild members: "Would Danny consider this a waste of time?" "How would Brian find something exciting here?" "Is this idea something Kevin or Brockman might run with, and where would they run with it?"

I seldom see my guild members in person (except the one I'm married to). We seldom talk on the phone. Yet we interact weekly through the crude old Internet tools of e-mail and links. (That no doubt reflects our age; younger guilds presumably use Facebook or Twitter or whatever's next in that lineage.)

Thanks to my guild's Internet-mediated conversation, my neuronal thinking is enhanced immeasurably by our digital thinking.

Trust Nothing, Debate Everything

Jason Calacanis

Internet entrepreneur; founder, Mahalo.com

As a former journalist, I used to withhold judgment and refrain from speculating about breaking stories until "all the facts" were in. I used to keep a mental scorecard of an issue with the confirmed facts neatly organized. However, with the velocity of information and tools to curate and process it on the Internet, I've moved to speculation as my scorecard. The real-time Web means we get to flip our positions, argue all sides of a debate, and test theories. We're being lied to and manipulated more than we're being told the truth, so instead of trying to figure out what's true, I'd rather speculate in my social network and see what comes back.

When the November 2009 shooting at Fort Hood happened, I immediately speculated on Facebook and Twitter that Nidal Malik Hasan's name was probably an indication of a terrorism link—it couldn't be coincidence, right? That was the first thing you thought, right? Dozens of responses came back, outraged that I would speculate to my 80,000 followers without "knowing for sure." Most claimed we should wait until the authorities completed their investigation. A couple of folks thought I was showing some bias against Muslims—which, of course, I was.

Any investigator would follow the radical Muslim pattern when faced with the same evidence, and certainly the newscasters on CNN were thinking it. The terrorism connection at Fort Hood was so obvious that the CNN reporters made a point of saying that just because the name sounded like the names of 9/11 hijackers, we shouldn't jump to conclusions. Really? Isn't that exactly what the investigators did? Isn't that what the Internet was doing while CNN anchors fumbled their way through the moment, trying to fill airtime with anything *but* speculation about radical Muslims?

They've tracked Hasan's connections to a mosque in Virginia where two of the September 11 hijackers attended services. Speculation on the Internet was correct this time, and CNN was doing the "responsible thing" by not participating in it. Really? Doesn't speculation lead to debate, which leads to, one hopes, some resolution?

Jumping to conclusions is a critical piece of information gathering, and we should be doing it more, not less. The Internet is built to route around bad routers and bad facts. Hasan's business card had "SoA" on it, which stands for—wait for it—"Soldier of Allah." If only someone had jumped to some conclusions about that fact on her Twitter account.

Consuming passive news gave way, in 2003 and 2004, to commenting on blogs. Now we all have blogs tethered to our mobile phones, even if they are micro in nature, with Facebook and Twitter accounts. We shouldn't wait for facts; we should be speculating and testing assumptions as news and knowledge unfold. Facts are, of course, valuable, but speculation gets me further and builds better webs in my mind.

We've moved from being jurors to being investigators, and the audience is onstage. Support thought bombs and the people who throw them into your social graph. It's messy but essential. Study the reactions on either side of the aisle, because reactions can be more telling than the facts sometimes. That's how the Internet has changed my thinking: Trust nothing, debate everything.

Harmful One-Liners, an Ocean of Facts, and Rewired Minds

Haim Harari

Physicist, former president, Weizmann Institute of Science; author, A View from the Eye of the Storm: Terror and Reason in the Middle East

It is entirely possible that the Internet is changing our way of thinking in more ways than I am willing to admit, but there are three clear changes that are palpable.

The first is the increasing brevity of messages.

Between Twittering, chatting, and sending abbreviated BlackBerry e-mails, the "old" sixty-second sound bite of TV newscasts is now converted into one-liners attempting to describe ideas, principles, events, complex situations, and moral positions. Even when the message itself is somewhat longer, the fact that we are exposed to more messages than ever before means that the attention dose allocated to each item is tiny. The result, for the general public, is a flourishing of extremist views on everything. Not just in politics, where only the ideas of the lunatic far left and far right can be stated in one sentence, but also in matters of science.

It is easy to state in one sentence nonsense such as "The theory of evolution is wrong," "Global warming is a myth," "Vaccinations cause autism," and "God"—mine, yours, or hers—"has all the answers." It requires long essays to explain and discuss the ifs and buts of real science and of real life.

I find that this trend makes me a fanatic antiextremist. I am boiling mad whenever I see or read such telegraphic (to use an ancient terminology) elaborations of ideas and facts, knowing that they are wrong and misleading yet find their way into so many hearts and minds. Even worse, people who are still interested in a deeper analysis and a balanced view of topics—whether scientific, social, political, or other—are considered leftovers from an earlier generation and labeled as extremists of the opposite color by the fanatics of one corner or another.

The second change is the diminishing role of factual knowledge in the thinking process. The thought patterns of different people on different subjects require varying degrees of knowing facts, being able to correlate them, creating new ideas, distinguishing between important and secondary matters, analyzing processes, knowing when to prefer pure logic and when to let common sense dominate, and numerous other components of a complex mental exercise.

The Internet allows us to know fewer facts, since we can be sure they are always literally at our fingertips, thus reducing their importance as a component of the thought process. This is similar to, but much more profound than, the reduced role of computation and simple arithmetic with the introduction of calculators. But we should not forget that often in the scientific discovery process the greatest challenges are to ask the right question rather than answer a well-posed question and to correlate facts that no one thought of connecting. The existence of many available facts somewhere in the infinite ocean of the Internet is no help in such an endeavor. I find that my scientific thinking is changed very little by the availability of all of these facts, though my grasp of social, economic, and political issues is enriched by having many more facts at my disposal. An important warning is necessary here: A crucial enhanced element of the thought process, demanded by the flood of available facts, must be the ability to evaluate the credibility of facts and quasi-facts. Both are abundant on the Web, and telling them apart is not as easy as it may seem.

The third change is in the entire process of teaching and learning. Here it is clear that the change must be profound and multifaceted, but it is equally clear that because of the ultraconservative nature of the educational system, change has not yet happened on a large scale. The Internet brings to us art treasures, the ability to simulate complex experiments, mechanisms of learning by trial and error, less need to memorize facts and numbers, explanations and lessons from the greatest teachers on Earth, special aids for children with special needs, and numerous other incomparable marvels unavailable to previous generations. Anyone involved in teaching, from kindergarten to graduate school, must be aware of the endless opportunities as well as the lurking dangers. These changes in learning, when they materialize, may create an entirely different pattern of knowledge, understanding, and thinking in the student mind.

I'm amazed by how little has changed in the world of education, but whether we like it or not, the change must happen, and it will happen. It may take another decade or two, but education will never be the same.

An interesting follow-up is the question of whether the minds and brains of children growing up in an Internet-inspired educational system will be physically "wired" differently from those of earlier generations. I tend to speculate in the affirmative, but this issue may be settled only by responses to the *Edge* question of 2040.

What Other People Think

Marti Hearst

Computer scientist, University of California, Berkeley, School of Information; author, Search User Interfaces

In graduate school, as a computer scientist whose focus was on search engines even before the Web, I always dreamed of an Internet that would replace the inefficiencies of libraries, making all important information easily available online. Amazingly, this came to pass, despite what seemed like insurmountable blockages in the early days.

But something I did not anticipate is how social the Internet would become. When the Web took off, I expected to see recipes online. But today I also expect to learn what other people thought about a recipe, including what ingredients they added, what salad they paired it with, and who in their family liked or disliked it. This multitude of perspectives has made me a better cook.

If I enjoy a television show, within minutes or hours of the airtime of the latest episode I expect to be able to take part in a delightful, informed conversation about it, anchored by an essay by a professional writer, supported with high-quality user-contributed comments that not only enhance my pleasure of the show but also reveal new insights.

And not only can I get software online, but in the last few years a dizzying cornucopia of free software components have appeared, making it possible to do in just days research and development that would have taken months or years in the past. There have always been online forums to discuss software—in fact, coding was one of the most common topics of early online groups. But the variety and detail of the kind of information that other people selflessly supply one another with today is staggering. And the design of online question-answering sites has moved from crufty to excellent in just a few years.

Most relevant to the scientists and researchers who contribute to the *Edge* question: We see the use of the Web to enhance communication in the virtual college, with academic meetings held online, math proofs done collaboratively on blogs, and deadly viruses isolated within weeks by research labs working together online.

Sure, we used e-mail in the early eighties, and there were online bulletin boards for at least a decade before the Web, but only a small percentage of the population used them, and it was usually over a very slow modem. In the early days of the Web, ordinary people's voices were limited primarily to information ghettos such as Geocities; most text was produced by academics and businesses. There was very little give-and-take. By contrast, according to a 2009 Pew study, 51 percent of Internet users now post online content that they have created themselves, and one in ten Americans posts something online every day.

Of course, increased participation means an increase in the equivalent of what we used to call flame wars, or generally rude behavior, as well as a proliferation of false information and gathering places for people to plan and encourage hurtful activities. Some people think this ruins the Web, but I disagree. It's what happens when everyone is there.

Interestingly, while the *Edge* question was innovative in format when it started, the Edge Website still does not allow readers to comment on the opinions offered. I'm not saying that this is either a good or a bad thing. The Edge Foundation's goal is to increase public understanding of science by encouraging intellectuals to "express their deepest thoughts in a manner accessible to the intelligent reading public." I just wonder whether it isn't time to embrace the new Internet and let that public write back.

The Extinction of Experience

Scott D. Sampson

Dinosaur paleontologist and science communicator; author, Dinosaur Odyssey: Fossil Threads in the Web of Life

Like that of many others, my personal experience is that the Internet is both the Great Source for information and the Great Distractor, fostering compulsions to stay connected, often at the expense of other, arguably more valuable aspects of life. I don't think the Internet alters the way I think as much as it does the way I work; having the Great Source close at hand is irresistible, and I generally keep a window open on my laptop for random searches that pop into my head.

Nevertheless, I'm much less concerned about "tweeners," who, like me, grew up before the Internet, than I am about children of the Internet age, the so-called digital natives. I want to know how the Internet changes the way *they* think. The jury is still out. Although the supporting research may still be years away, it seems likely that a lifetime of daily conditioning dictated by the rapid flow of information across glowing screens will generate substantial changes in brains and thus in thinking. Commonly cited potential effects include fragmented thinking and shorter attention spans, with a concomitant reduction in reflection (let alone interest), introspection, and in-depth thought. Another oft-noted concern is the nature of our communications, which are becoming increasingly terse and decreasingly face-to-face.

But I have a larger fear, one rarely mentioned in these discussions—the extinction of experience. This term, which comes from author and naturalist Robert Michael Pyle, refers to the loss of intimate experience with the natural world. Clearly, anyone who spends ten-plus hours each day focused on a screen is not devoting much time to experiencing the "real" world. More and more, it seems, real-life experience is being replaced by virtual alternatives. And—to my mind, at least—this is a grave problem.

As the first generation to contemplate the fact that humanity may have a severely truncated future, we live at arguably the most pivotal moment in the substantial history of *Homo sapiens*. Decisions made and actions taken during the next generation will have an imbalanced effect on the future of humans and all other life on Earth. If we blunder onward on our present course—increasing populations, poverty, greenhouse gas emissions, and habitat destruction—we face no less than the collapse of civilization and the destruction of the biosphere. Given the present dire circumstances, any new far-reaching cultural phenomenon must be evaluated in terms of its ability to help or hinder the pressing work to be done; certainly this concern applies to how the Internet influences thinking.

Ecological sustainability, if it is to occur, will include greener technologies and lifestyles. In addition, however, we require a shift in worldview that reconfigures our relationship with nonhuman nature. To give one prominent example of our current dysfunctional perspective, how are we to achieve sustainability as long as we see nature as part of the economy rather than the inverse? Instead of being a collection of resources available for our exploitation, nature must become a community of relatives worthy of our respect and a teacher to whom we look for inspiration and insight. In contrast to the present day, sustainable societies will likely be founded on local foods, local materials, and local energy. They will be run by people who have a strong passion for place and a deep understanding of the needs of those places. And I see no way around the fact that this passion and understanding will be grounded in direct, firsthand experiences with those places.

My concern, then, is this: How are we to develop new, more meaningful connections to our native communities if we are staring at computer screens that connect us only to an amorphous worldwide "community"? As is evident to anyone who has stood in a forest or on a seashore, there is a stark difference between a photograph or video and the real thing. Yes, I understand the great potential for the Internet to facilitate fact-finding, information sharing, and even community building by like-minded people. I am also struck by the radical democratization of information that the Internet may soon embody. But how are we to establish affective bonds locally if our lives are consumed by virtual experiences on global intermedia? What we need is uninterrupted solitude outdoors, sufficient time for the local sights, sounds, scents, tastes, and textures to seep into our consciousness. What we are seeing is children spending less and less time outdoors experiencing the real world and more and more time indoors immersed in virtual worlds.

In effect, my argument is that the Internet may influence thinking indirectly through its unrelenting stranglehold on our attention and the resultant death (or at least denudation) of nonvirtual experience. If we are to care about larger issues surrounding sustainability, we first must care about our local places, which in turn necessitates direct experiences in those places. As Pyle observes, "What is the extinction of the condor to a child who has never known a wren?"

One thing is certain: We have little time to get our act together. Nature, as they say, bats last. Ultimately, I can envision the Internet as a net positive or a net negative force in the critical sustainability effort, but I see no way around the fact that any positive outcome will involve us turning off the screens and spending significant time outside, interacting with the real world—in particular, the nonhuman world.

The Collective Nature of Human Intelligence

Matt Ridley

Science writer; founding chairman, International Centre for Life; author, The Rational Optimist: How Prosperity Evolves

The Internet is the ultimate mating ground for ideas, the supreme lekking arena for memes. Cultural and intellectual evolution depends on sex just as much as biological evolution does; otherwise they remain merely vertical transmission systems. Sex allows creatures to draw upon mutations that happen anywhere in their species. The Internet allows people to draw upon ideas that occur to anybody in the world. Radio and printing did this, too, and so did writing and before that language, but the Internet has made it fast and furious.

Exchange and specialization are what makes cultural and intellectual evolution happen, and the Internet's capacity for encouraging exchange encourages specialization too. Somebody somewhere knows the answer to any question I care to ask, and it is much easier to find him or her. Often it is an amateur, outside journalism or academia, who just happens to have a piece of knowledge at hand. An example: Suspicious of the claim that warm seas (as opposed to rapidly warming seas) would kill off coral reefs, I surfed the Net until I found the answer to the question, Is there any part of the oceans that is too hot for corals to grow? One answer lay in a blog comment from a diver just back from the Iranian side of the Persian Gulf, where he had seen diverse and flourishing coral reefs in 35°C water (10 degrees warmer than much of the Great Barrier Reef).

This has changed the way I think about human intelligence. I've never had much time for the academic obsession with intelligence. Highly intelligent people are sometimes remarkably stupid; stupid people sometimes make better leaders than clever ones. And so on.

The reason, I realize, is that human intelligence is a collective phenomenon. If they exchange and specialize, a group of fifty dull-witted people can have a far higher collective intelligence than fifty brilliant people who don't. That's why it is utterly irrelevant if one race turns out to have higher IQ than another, or one company hires more people with higher IQs than another. I would rather be marooned on a desert island with a diverse group of mediocre people who know how to communicate than with a bunch of geniuses. The Internet is the latest and best expression of the collective nature of human intelligence.

Six Ways the Internet May Save Civilization

David Eagleman

Neuroscientist; novelist; director, Laboratory for Perception and Action, Baylor College of Medicine; author, Sum

The Internet has changed the way I think about the threat of societal collapse. When we learn of the empires that have tumbled before us, it is plausible to think that our civilization will follow the same path and eventually fall to a traditional malady—anything from epidemics to resource depletion. But the rapid advance of the Internet has thoroughly (and happily) changed my opinion about our customary existential threats. Here are six ways that the possession of a rapid and vast communication network will make us much luckier than our predecessors:

1. Sidelining Epidemics

One of the more dire prospects for collapse is an infectious disease epidemic. Bacterial or viral epidemics precipitated the fall of the golden age of Athens, the Roman Empire, and most empires of the Native Americans. The Internet can be our key to survival, because the ability to work telepresently can inhibit microbial transmission by reducing human-to-human contact. In the face of an otherwise devastating epidemic, businesses can keep supply chains running with the maximum number of employees working from home. This won't keep everyone off the streets, but it can reduce host density below the tipping point. If we are well prepared when an epidemic arrives, we can fluidly shift into a self-quarantined society in which microbes fail due to host sparseness. Whatever the social ills of isolation, they bode worse for the microbes than for us.

2. Availability of Knowledge

Important discoveries have historically stayed local. Consider smallpox inoculation. This practice was under way in India, China, and Africa for at least a hundred years before it made its way to Europe. By the time the idea reached North America, the native civilizations had long collapsed. And information is not only hard to share but also hard to keep alive. Collections of learning from the library at Alexandria to the Mayan corpus have fallen to the bonfires of invaders or the winds of natural disaster. Knowledge is hard won but easily lost.

The Internet addresses the problem of knowledge sharing better than any technology we've had. New discoveries latch on immediately: The information spreads widely and the redundancy prevents erasure. In this way, societies can use the latest bricks of knowledge in their fortification against existential threats.

3. Speed by Decentralization

We are witnessing the downfall of slow central control in the media: News stories are increasingly becoming user-generated nets of dynamically updated information. During the recent California wildfires, locals went to the TV stations to learn whether their neighborhoods were in danger. But the news stations appeared most concerned with the fate of celebrity mansions, so Californians changed their tack: They tweeted, uploaded geotagged cell phone pics, and updated Facebook. The Internet carried the news more quickly and accurately than any news station could. In this decentralized regime, there were embedded reporters on every neighborhood block, and the news shock wave kept ahead of the fire front. In the right circumstances, this head start could provide the extra hours that save us.

4. Minimization of Censorship

Political censorship has been a familiar specter in the last century, with state-approved news outlets ruling the press, airwaves, and copying machines in the former USSR, Romania, Cuba, China, Iraq, and other countries. In all these cases, censorship hobbled the society and fomented revolutions. Historically, a more successful strategy has been to confront free speech with free speech—and the Internet allows this in a natural way. It democratizes the flow of information by offering access to the newspapers of the world, the photographers of every nation, the bloggers of every political stripe. Some postings are full of doctoring and dishonesty, others strive for independence and impartiality, but all are available for the end user to sift through for reasoned consideration.

5. Democratization of Education

Most of the world does not have access to the education afforded to a small minority. For every Albert Einstein, Yo-Yo Ma, or Barack Obama who has the opportunity for education, there are uncountable others who never get the chance. This vast squandering of talent translates directly into reduced economic output. In a world where economic meltdown is often tied to collapse, societies are well advised to leverage all the human capital they have. The Internet opens the gates of education to anyone who can get her hands on a computer. This is not always a trivial task, but the mere feasibility redefines the playing field. A motivated teen anywhere on the planet can walk through the world's knowledge, from Wikipedia to the curricula of MIT's OpenCourseWare.

6. Energy Savings

It is sometimes argued that societal collapse can be cast in terms of energy: When energy expenditure begins to outweigh energy return, collapse ensues. The Internet addresses the energy problem with a kind of natural ease. Consider the huge energy savings inherent in the shift from snail mail to e-mail. As recently as the last decade, information was amassed not in gigabytes but in cubic meters of filing cabinets. Beyond convenience, it may be that the technological shift from paper to electrons is critical to the future. Of course, there are energy costs to the banks of computers that underpin the Internet, but these costs are far less than the forests and coal beds and oil deposits that would be depleted for the same quantity of information flow.

The tangle of events that trigger societal collapse can be complex, and there are several existential threats the Internet does not address. Nonetheless, it appears that vast, networked communication can serve as an antidote to several of the most common and fatal diseases of civilization. Almost by accident, we now command the capacity for self-quarantining, retaining knowledge, speeding information flow, reducing censorship, actualizing human capital, and saving energy resources. So the next time a co-worker laments about Internet addiction, the banality of tweets, or the decline of face-to-face conversation, I will sanguinely suggest that the Internet—even with all its flashy wastefulness—just might be the technology that saves us.

Better Neuroxing Through the Internet

Samuel Barondes

Director of the Center for Neurobiology and Psychiatry at the University of California, San Francisco; author, Better Than Prozac: Creating the Next Generation of Psychiatric Drugs

Years ago, when Xerox machines first came to libraries, many of us breathed a sigh of relief. Instead of copying passages from journals in our barely legible script, we could put the important pages on the scanner and print a good replica that we could turn to whenever we liked. The process soon became so cheap that we could duplicate whole articles we were interested in, and then even articles we *might* be interested in. Soon we had piles of this stuff wherever we turned.

The biologist Sydney Brenner, who likes to pepper his research with observations about human folly, quickly realized that this technology also provided new opportunities for wasting time, because many people were photocopying and filing a lot of irrelevant papers instead of carefully reading and remembering the key points of the most significant ones. This led to his playful warning that it is much more important to be "neuroxing" than Xeroxing.

Brenner's famous caveat didn't do much to shorten the copier lines. But the Internet did. Instead of collecting reprints by feeding the machine one page at a time, the Internet allows us to build personal libraries of PDFs just by clicking on links. It also allows us to keep up to date on the matters we are especially interested in by setting up alerts and to keep sampling new fields in as much depth as we choose.

And the good news is that by eliminating our reliance on libraries and copiers while instantaneously providing user-friendly access to information, this new technology is clearly facilitating intellectual activities rather than getting in their way. This is not to say that the Internet is free of time-wasting temptations. But if you want the latest and most relevant data about whatever you are interested in, the Internet can bring much of it to you in the blink of an eye. All ready to be neuroxed.

A Gift to Conspirators and Terrorists Everywhere

Marcel Kinsbourne

Neurologist and cognitive neuroscientist, the New School; author, Children's Learning and Attention Problems

The Internet has supplied me with an answer to a question that has exercised me interminably: When I reach Heaven (surely!), how can I possibly spend infinite time without incurring infinite boredom? Well, as long as they provide an Internet connection, I now see that I can.

The instant response, correct or otherwise, to every question sets up an intellectual Ponzi scheme. The answer multiplies the questions, which, in a potentially infinite progress, prompt yet more. Having previously functioned in serial fashion, digging for new vistas through largely unexplored libraries, I now have to neither interact with any other human being nor even move, except my fingers. And I can pursue as many ideas in parallel as I want. The only hitch is the desire of various business concerns to make me pay for the information I crave. Once we have reached the stage of having universal Internet access implanted in our brains, even that will no longer be a problem, because we will be dealing in thoughts, and thoughts are famously considered to be free, if not gratuitous.

Multiplied by a multitude, and compounded over time, this proliferation of ideas will offer a potential for stacking invention on invention, scaling up to accomplishments undreamed by science fiction. And that will be just as well, because of the countervailing menace of the Internet. Here's why.

Evolution, generally a good thing, comes with two intractable problems: It is excruciatingly slow, and it is totally lacking in foresight (that is, its design is seriously unintelligent). The progeny lives with the unforeseen consequences.

Our near-human ancestors were scattered in small groups across inhospitable and predator-infested savanna. Individuals ruthless and cruel enough to repel those competing for scarce resources were favored by natural selection, did their thing, and the species survived. Their inborn fury knew no bounds, which was not then much of a problem for external reasons: the bounds set on their ability to destroy by the short range of their weaponry (clubs), their sluggish transportation (legs), and their feeble vehicle of communication (voice), unable to reach outside their band to conjoin with others similarly inclined so as to wreak havoc in substantial numbers. But as cultural innovation outstripped evolution with exponential momentum, the means for harm gained efficacy. In the meantime, the destroyers persisted as small minorities in every population group, although as resources have become less scarce, their assistance to help the group survive has (or should have) become less and less needed.

Advances in weaponry have brought us to the point of being able to deliver havoc to all parts of Earth and at great speed. Only communication lagged behind in recent years, though radio and television did begin to infiltrate and integrate greater masses of the population. Yet the human population grew and grew, despite the massacre of multitudes by those so inclined. It seemed for a shining moment in history that the constructive was outstripping destructive. But nothing succeeds as planned.

How to coordinate limited numbers of like-minded destroyers the world over, so as collectively to inflict maximum harm? Use the Internet, Web networks, to recruit and plan; it is a gift to conspirators and terrorists everywhere. The pace of the arms race is accelerating, while evolution is left way behind. Terror becomes globalized, and through it the prospect of global suicide. Why would human beings want that? Because it is in their nature.

Consider the frog and the scorpion. Give me a ride across the stream. But you will sting me. and I will die, replies the frog. But then I would drown, argues the scorpion. The frog swims, carrying his passenger, feels an ominous sting. Why? he asks. Because it is my nature, replies the scorpion.

Natural selection selects but cannot explain why. After all, there is no one there to explain. So those naturally selected act according to their nature, then and now (but now with far greater reach). Blame it on unintelligent design.

There is a dynamic of cumulative invention in the human brain. A dynamic of insensate destruction is also inherent in the human brain. Behold the ultimate great arms race, brought to a head by the Internet, which acts as a double agent, aiding and energizing both sides. Will it perfect the species or drive it into extinction? The cockroaches will bear witness.

The Ant Hill

Eva Wisten

Journalist; author, Single in Manhattan

When you're on a plane, watching the cars below and the blinking, moving workings of a city, it's easy to believe that everything is connected—just moving parts in the same system. If you're one of the drivers on the ground, driving your car from B to A, the perspective is, of course, different. The driver feels very much like an individual—car to match your personality, on the way to your chosen destination. The driver never feels like a moving dot in a row of a very large number of other moving dots.

The Internet sometimes makes me suspect I'm that driver. The merging (often invisible) of many disparate systems is steering my behavior into all kinds of paths, which I can only hope are beneficial. The visible connectedness through the Web has maybe not changed how I think but has increased the number of people whose thoughts are in my head. Because of the Internet, the memes and calculations of more and more people (and/or computers) pass through us. Good or bad, this new level of connectedness sometimes makes me feel that if I could only hover a few feet off the ground, what I would see is an ant hill. All the ants, looking so different and special up close, seem suspiciously alike from this height. This new tool for connections has made more ants available every time I need to transport a twig, just as there are more ants in the way when I want to set down the picnic basket.

But—as a larger variety of thoughts and images pass by, as I search a thought and see the number of people who have had the same thought before me, as more and more systems talk to one another and take care of all kinds of logistics—I do think this level of connectedness has pushed us, beneficially, toward both the original and the local.

We can go original, either in creation or curation, and carve a new little path in the ant hill, or we can copy one of all the things out there and bring it home to our local group. Some ants manage to be original enough to benefit the whole anthill, but other ants can copy and modify the good stuff and bring it home. And in this marching back and forth, trying to get things done, communicate, make sense of things, I look not to leaders but to curators, who can efficiently signal where to find the good stuff.

What is made accessible to me through the Internet might not be changing how I think, but it does some of my thinking for me. Above all, the Internet is changing how I see myself. As real-world activity and connections continue to be what matters most to me, the Internet, with its ability to record my behavior, is making it clearer that I am, in thought and in action, the sum of the thoughts and actions of other people to a greater extent than I have realized.

I Can Make a Difference Because of the Internet

Bruce Hood

Director, Bristol Cognitive Development Centre, Experimental Psychology Department, University of Bristol, U.K.; author, SuperSense: Why We Believe in the Unbelievable

Who has not Googled himself? Most humans have a concept of self constructed in terms of how we think we're perceived by those around us, and the Internet has made that preoccupation trivially easy. Now people can assess their impact factor through a multitude of platforms, including Facebook, Twitter, and, of course, blogging.

Last year, at the request of my publisher, I started a blog to comment on weird and bizarre examples of supernatural thinking from around the world. At the outset, I thought blogging was a self-indulgent activity, but I agreed to give it a whirl to help promote my book. In spite of my initial reluctance, I soon became addicted to feedback. It was not enough to post blogs for some unseen audience. I needed validation from visitors that my efforts and opinions were appreciated. Within weeks, I had become a numbers junkie looking for more and more hits.

However, the Internet has also made me aware of both my insignificance and my power. Within the blogosphere, I am no longer an expert on any opinion, since it can be shared or rejected by a multitude of others. But insignificant individuals can make a significant difference when they coalesce around a cause. As I write this, a British company is under public scrutiny for allegedly selling bogus bomb-detecting dowsing rods to the Iraqi security forces. This scrutiny has come about because of a blog campaign by like-minded skeptics, who used the Internet to draw attention to what they considered to be questionable business activity. Such a campaign would have been difficult in the pre-Internet days and not something that the ordinary man in the street would have taken on. In this way, the Internet empowers the individual. I can make a difference because of the Internet. I'll be checking back on Google to see if anyone shares my opinion.

Go Virtual, Young Man

Eric Weinstein

Mathematician and economist; principal, Natron Group

Oddly, the Internet is still invisible, to the point where many serious thinkers continue to doubt whether it changes modern thought at all.

In science, we generally first learn about invisible structures from anomalies in concrete systems. The existence of an invisible neutrino on the same footing as visible particles was predicted in 1930 by Wolfgang Pauli as the error term necessary to save the principles of conservation of energy and momentum in beta decay. Likewise, human memes invisible to DNA (e.g., tunes) were proposed in 1976 by Richard Dawkins, since selection, to remain valid, must include all self-replicating units of transmission involved in tradeoffs with traditional genes.

Following this line of thinking, it is possible that a generalized Internet may even be definable, with sufficient care, as a kind of failure of the physical world to close as a self-contained system. Were a modern Rip van Winkle sufficiently clever, he might eventually infer something like the existence of file-sharing networks from witnessing the collapse of music stores, CD sales, and the recording industry's revenue model.

The most important example of this principle has to do with markets and geography. The Internet has forced me to view physical and intellectual geography as instances of an overarching abstraction coexisting on a common footing. As exploration and trade in traditional physical goods such as spices, silk, and gold have long been linked, it is perhaps unsurprising that the marketplace of ideas should carry with it an intellectual geography all its own. The cartography of what may be termed the Old World of ideas is well developed. Journals, prizes, and endowed chairs give us landmarks to which we turn in the quest for designated thinkers, and for those wishing to hug the shore of the familiar, this proves a great aid.

Despite being relatively stable, the center of this scientific world began to shift in the last century from institutions in Europe to ones in North America. While there is currently a great deal of talk about a second shift, from the United States toward Asia, it may instead happen that the next great migration will be dominated by flight to structures in the virtual from those moored to the physical.

Consider the award in 2006 of the Fields Medal (the highest prize in mathematics) for a solution of the Poincaré Conjecture. This was remarkable in that the research being recognized was not submitted to any journal. In choosing to decline the medal, peer review, publication, and employment, the previously obscure Grigori Perelman chose to entrust the legacy of his great triumph solely to an Internet archive intended as a temporary holding tank for papers awaiting publication in established journals. In so doing, he forced the recognition of a new reality by showing that it was possible to move an indisputable intellectual achievement out of the tradition of referee-gated journals bound to the stacks of university libraries into a new and poorly charted virtual sphere of the intellect.

But while markets may drive exploration, the actual settlement of the frontier at times requires the commitment of individuals questing for personal freedom, and here the New World of the Internet shines. It is widely assumed that my generation failed to produce towering figures like Francis Crick, P. A. M. Dirac, Alexander Grothendieck, or Paul Samuelson because something in the nature of science had changed. I do not subscribe to that theory. Suffice it to say that issues of academic freedom have me longing to settle among the noble homesteaders now gathering on the efficient frontier of the marketplace of ideas. My intellectual suitcases have been packed for months now, as I try to screw up the courage and proper "efficient frontier mentality" to follow my own advice to the next generation: "Go virtual, young man."

My Internet Mind

Thomas A. Bass

Professor of English, University at Albany, State University of New York; author, The Spy Who Loved Us

What do I do with the Internet? I send out manuscripts and mail, buy things, listen to music, read books, hunt up information and news. The Internet is a great stew of opinion and facts. It is an encyclopedic marvel that has transformed my world. It has also undoubtedly transformed the way I think.

But if we humans are the sex organs of our technologies, reproducing them, expanding their domains and functionality—as Marshall McLuhan said—then perhaps I should turn the question upside down. Because of my reliance on the Internet, the number of hours each day I spend in its electronic embrace, have I begun to think like the Internet? Do I have an Internet mind that has been transformed by my proximity to this network of networks?

How does the Internet think? What does it want of me, as I go about distractedly meeting its demands? Again to cite McLuhan, this time quoting in full the passage that describes my outed brain and airborne nerves:

> Electronic technology requires utter human docility and quiescence of meditation such as befits an organism that now wears its brain outside its skull and its nerves outside its hide. Man must serve his electric technology with the same servo-mechanistic fidelity with which he served his coracle, his canoe, his typography, and all the other extensions of his physical organs.

I have used the word *distracted* to describe my Internet mind. We all know the feeling of being jumpy, edgy, nervous around the Net. Time is speeding up. Space is contracting. Sentences are getting shorter. Thoughts are swifter—dare we say shallower? Again, McLuhan got the jump on us. Fifty years ago, he announced, "Mental breakdown of varying degrees is the very common result of uprooting and inundation with new information and endless new patterns of information." Our "electric implosion" has ushered in an "age of anxiety."

This distracted state will end, said McLuhan, when our machines begin to think on their own. They will be smarter than us, as they already are in lots of ways, such as calculating numbers and flying airplanes. "Having extended or translated our central nervous system into the electromagnetic technology, it is but a further stage to transfer our consciousness to the computer world as well." This final handoff from man to machine will allow us to "program consciousness," said McLuhan.

Luckily for those of us who, when we check the headlines, sometimes find our mouse hovering over a picture of the latest celebrity scandal, the computer consciousness currently evolving beyond our human minds will be more dignified. McLuhan assured us that this new consciousness would be free of "the Narcissus illusions of the entertainment world that beset mankind when he encounters himself extended in his own gimmickry."

A Catholic mystic touched by the spiritual optimism of Teilhard de Chardin, McLuhan foresaw a glorious end to my acquisition of an Internet mind. "The computer, in short, promises by technology a Pentecostal condition of universal understanding and unity," he said. With computers functioning as translating machines, allowing me "to by-pass languages in favor of a general cosmic consciousness," I will be ushered eventually into "a perpetuity of collective harmony and peace." In the meantime, excuse me while I go check the headlines, pay some bills, and try not to click on too many of today's top-ten distractions.

"If You Have Cancer, Don't Go on the Internet"

Karl Sabbagh

Writer and television producer; author, Remembering Our Childhood: How Memory Betrays Us

When the British playwright Harold Pinter developed cancer of the esophagus, his wife, Lady Antonia Fraser, discovered online that there was a 92 percent mortality rate. "If you have cancer, don't go on the Internet," she told the *Sunday Times* in January 2010.

This set me thinking about my own interactions with the Internet and how they might differ fundamentally from my use of other sources of information.

Lady Antonia could, I suppose, have said, "If you have cancer, don't look at the *Merck Manual*," or some other medical guide, but there must be more to it than that. First of all, there's the effortlessness of going on the Internet. I used to joke that if I had a query that could be answered either by consulting a book in the shelves on the other side of my study or by using the Internet, it would be quicker and less energy-consuming to find the answer on the Internet. That's not funny anymore, because it's obviously the most efficient way to do things. I'm one of the few people who trust Wikipedia; its science entries, in particular, are extremely thorough, reliable, and well sourced. People who trust books (two or more years out of date) rather than Wikipedia are like people who balk at buying on the Internet for security reasons but happily pay with a credit card in restaurants, where an unscrupulous waiter can keep the carbon copy of your slip and run up huge bills before you know it.

Lady Antonia's remark was really a tribute to the reliability and comprehensiveness of the Internet. Probably it wasn't so much that she came across a pessimistic prognosis as that she came across a presumably *reliable* one, based on up-to-date information. That doesn't, of course, mean it was accurate. She may not have consulted all cancer sites, or it may be that no one really knows for sure what the prognosis is for esophageal cancer. But she assumed—and I, too, assume when using the Internet—that with a little skill and judgment you can get more reliable information there than anywhere else.

This, of course, has nothing to do with thinking. It could be that I would think the same if I'd been writing my books with a quill pen and had only the Bible, Shakespeare, and Dr. Johnson's *Dictionary* to consult. But the Internet certainly constrains what I think about. It stops me from thinking any more about that great idea for a book that I now find was published a few years ago by a small university press in Montana.

It also reinforces my belief in my own ideas and opinions, because it is now much quicker to test them, particularly when they are new opinions. By permitting anyone to publish anything, the Internet allows me to read the whole range of views on a topic and infer from the language used the reasonableness or otherwise of the views. Of course, I was inclined to disbelieve in Intelligent Design before I had access to the wide range of wacky and hysterical Websites that promote it. But now I have no doubts at all that the theory is tosh (SLANG, CHIEFLY BRIT nonsense; rubbish —*The Free Dictionary*).

But this is still not to do with *thinking*. What do I do all day, sitting at my computer? I string words together, reread them, judge them, improve them if necessary, and print them out or send them to people. And underlying this process is a judgment about what is interesting, novel, or in need of explanation—and the juggling of words in my mind to express these concepts in a clear way. None of that, as far as I am aware, has changed because of the Internet.

But this is to deal with only one aspect of the Internet: its provision of factual content. There are also e-mail and attachments and blogs and software downloads and YouTube and Facebook and shopping and banking and weather forecasts and Google Maps and and and . . . But before all this, I knew there were lots of people in the world capable of using language and saying clever or stupid things. Now I have access to them in a way I didn't before—but, again, this is just information provision rather than a change in ways of thinking.

Perhaps the crucial factor is speed. When I was setting out to write a book, I would start with a broad outline and a chapter breakdown, and these would lead me to set a series of research tasks that could take months: look in this library, write to this expert, look for this book, find this document. Now the order of things has changed. While I was doing all the above, which could take weeks or months, my general ideas for the book would be evolving. My objectives might change and my research tasks with them. I would do more broad-brush thinking. Now, when documents can be found and downloaded in seconds, library catalogs consulted from one's desk, experts e-mailed, and a reply received within twenty-four hours, the idea is set in stone much earlier.

But even here, there is no significant difference in thinking. If in the course of the research some document reveals a different angle, that this happens within hours or days rather than months can only be to the good. The broad-brush thinking is now informed rather than uninformed.

I give up. The Internet *hasn't* changed how I think. It's only a tool. An electric drill wouldn't change how many holes I make in a piece of wood; it would only make the hole-drilling easier and quicker. A car doesn't change the nature and purpose of a journey I make to the nearest town; it only makes it quicker and leads to me making more journeys than if I walked.

But what about Lady Antonia Fraser? Is the truth-telling power of the Internet something to avoid? The fact is, the Internet reveals in its full horror the true nature of humankind—its obsessions, the triviality of its interests, its scorn for logic or rationality, its inhumanity, the power of capital, the intolerance of the other. But anyone who says this is news just doesn't get out enough. The Internet magnifies and specifies what we know already about human nature—or if we don't know it, we're naïve. The only way my thinking could have been changed by this "revelation" would have been if I believed, along with Dr. Pangloss, that all is for the best in the best of all possible worlds. And I don't.

Incomprehensible Visitors from the Technological Future

Alison Gopnik

Psychologist, University of California, Berkeley; author, The Philosophical Baby: What Children's Minds Tell Us About Truth, Love, and the Meaning of Life

My thinking has certainly been transformed in alarming ways by a relatively recent information technology, but it's not the Internet. I often sit for hours in the grip of this compelling medium, motionless and oblivious, instead of interacting with the people around me. As I walk through the streets, I compulsively check out even trivial messages (movie ads, street signs), and I pay more attention to descriptions of the world (museum captions, menus) than to the world itself. I've become incapable of using attention and memory in ways that previous generations took for granted.

Yes, I know, reading has given me a powerful new source of information. But is it worth the isolation, or the damage to dialog and memorization that Socrates foresaw? Studies show, in fact, that I've become involuntarily *compelled* to read; I can't keep myself from decoding letters. Reading has even reshaped my brain: Cortical areas that once were devoted to vision and speech have been hijacked by print. Instead of learning through practice and apprenticeship, I've become dependent on lectures and textbooks. And look at the toll of dyslexia and attention disorders and learning disabilities—all signs that our brains were not designed to deal with such a profoundly unnatural technology.

Like many others, I suspect that the Internet has made my experience more fragmented, splintered, and discontinuous. But I'd argue that that's not because of the Internet itself but because I have mastered the Internet as an adult. Why don't we feel the same way about reading and schooling that we feel about the Web? Those changes in the way we get information have had a pervasive and transformative effect on human cognition and thought, and universal literacy and education have been around for only a hundred years or so.

It's because human change takes place across generations rather than within a single life. This is built into the very nature of the developing mind and brain. All the authors of these essays have learned how to use the Web with brains that were fully developed long before we sent our first e-mail. All of us learned to read with the open and flexible brains we had when we were children. As a result, no one living now will experience the digital world in the spontaneous and unself-conscious way that the children of 2010 will experience it, or in the spontaneous and unself-conscious way we experience print.

There is a profound difference between the way children and adults learn. Young brains are capable of much more extensive change—more rewiring—than the brains of adults. This difference between old brains and young ones is the engine of technological and cultural innovation. Human adults, more than any other animal, reshape the world around them. But adults innovate slowly, intentionally, and consciously. The changes that take place within an adult life, like the development of the Internet, are disruptive, attention-getting, disturbing, or exciting. But those changes become second nature to the next generation of children. Those young brains painlessly absorb the world their parents created, and that world takes on a glow of timelessness and eternity, even if it was created only the day before you were born.

My experience of the Web feels fragmented, discontinuous, and effortful (and interesting) because, for adults, learning a new technology depends on conscious, attentive, intentional processing. In adults, this kind of conscious attention is a limited resource. This is true even at the neural level. When we pay attention to something, the prefrontal cortex—the part of our brain responsible for conscious, goal-directed planning—controls the release of cholinergic transmitters, chemicals that help us learn, to certain very specific parts of the brain. So as we wrestle with a new technology, we adults can change our minds only a little bit at a time.

Attention and learning work very differently in young brains. Young animals have much more widespread cholinergic transmitters than adults do, and their ability to learn doesn't depend on planned, deliberate attention. Young brains are designed to learn from everything new or surprising or information-rich, even when it isn't particularly relevant or useful.

So children who grow up with the Web will master it in a way that will feel as whole and natural as reading feels to us. But that doesn't mean their experience and attention won't be changed by the Internet, any more than my print-soaked twentieth-century life was the same as the life of a barely literate nineteenth-century farmer.

The special attentional strategies we require for literacy and schooling may feel natural because they are so pervasive, and because we learned them at such an early age. But at different times and places, different ways of deploying attention have been equally valuable and felt equally natural. Children in Mayan Indian cultures, for example, are taught to distribute their attention to several events simultaneously, just as print and school teach us to focus on only one thing at a time. I'll never be able to deploy the broad yet vigilant attention of a hunter-gatherer—though, luckily, a childhood full of practice caregiving let me master the equally ancient art of attending to work and babies at the same time.

Perhaps our digital grandchildren will view a master reader with the same nostalgic awe we now accord to a master hunter or an even more masterly mother of six. The skills of the hyperliterate twentieth century may well disappear—or, at least, become highly specialized enthusiasms, like the once universal skills of hunting, poetry, and dance. It is sad that after the intimacy of infancy our children inevitably end up being somewhat weird and incomprehensible visitors from the technological future. But the hopeful thought is that my grandchildren will not have the fragmented, distracted, alienated digital experience that I do. To them, the Internet will feel as fundamental, as rooted, as timeless, as a battered Penguin paperback, that apex of the literate civilization of the last century, feels to me.

"Go Native"

Howard Gardner

Psychologist, Harvard University; author, Changing Minds

The Internet has changed my life greatly, but not in a way that I could have anticipated, nor in the way that the question implies. Put succinctly, just as if a newly discovered preliterate tribe had challenged my beliefs about human language and human culture, the Internet has altered my views of human development and human potential.

Several years ago, I had a chance conversation with Jonathan Fanton, then president of the MacArthur Foundation. He mentioned that the foundation was sponsoring, to the tune of $50 million, a major study of how young people are being changed by the new digital media, such as the Internet. At the time, as part of the GoodWork research project, I was involved in studies of ethics, focusing particularly on the ethical orientation of young people. So I asked Fanton, "Are you looking at the ways in which the ethics of youth may be affected?" He told me that the foundation had not thought about this issue. After several conversations and a grant application, my colleagues and I launched the GoodPlay project, a social science study of ethics in the digital media.

Even though I myself am a digital immigrant—I sometimes refer to myself as a digital paleolith—I now spend many hours a week thinking about the ways in which nearly all of us, young and old, are affected by being online, networked, and surfing or posting for so much of the day. I've become convinced that the "digital revolution" may be as epoch-making as the invention of writing or, certainly, the invention of printing or of broadcasting. While I agree with those who caution that it is premature to detail the effects, it is not too early to begin to think, observe, reflect, or conduct pivotal observations and experiments. Indeed, I wish that social scientists and/or other observers had been around when earlier new media debuted.

Asked for my current thinking, I would make the following points. The lives and minds of young people are far more fragmented than at earlier times. This multiplicity of connections, networks, avatars, messages, may not bother them but certainly makes for identities that are more fluid and less stable. Times for reflection, introspection, solitude, are scarce. Long-standing views of privacy and ownership/authorship are being rapidly undermined. Probably most dramatically, what it has meant for millennia to belong to a community is being totally renegotiated as a result of instant, 24/7 access to anyone connected to the Internet. How this will affect intimacy, imagination, democracy, social action, citizenship, and other staples of humankind is up for grabs.

For older people (even older than I am), the digital world is mysterious. For those of us who are middle-aged or beyond, we continue to live in two worlds—the predigital and the digital—and we may either be nostalgic for the days without BlackBerrys or relieved that we no longer have to trudge off to the library. But all persons who want to understand their children or their grandchildren must make the effort to "go native"—and at such times we digital immigrants or digital paleoliths can feel as fragmented, as uncertain about privacy, as pulled by membership in diverse and perhaps incommensurate communities, as any fifteen-year-old.

The Maximization of Neoteny

Jaron Lanier

Musician, computer scientist; pioneer of virtual reality; author, You Are Not a Gadget: A Manifesto

The Internet, as it evolved up to about the turn of the century, was a great relief and comfort to me and influenced my thinking positively in a multitude of ways. There were the long-anticipated quotidian delights of speedy information access and transfer, but also the far more important optimism born from seeing so many people decide to create Web pages and become expressive—proof that the late twentieth century's passive society on the couch in front of the TV was only a passing bad dream.

In the last decade, however, the Internet has taken on unpleasant qualities and has become gripped by reality-denying ideology.

The current mainstream dominant culture of the Internet is the descendant of what used to be the radical culture of the early Internet. The ideas, unfortunately, are motivated to a significant degree by a denial of the biological nature of personhood. The new true believers attempt to conceive of themselves as becoming ever more like abstract, immortal, information machines instead of messy, mortal, embodied creatures. This is just another approach to an ancient folly—the psychological denial of aging and dying. To be a biological realist today is to hold a minority opinion in an age of profound, overbearing, technologically enriched groupthink.

When I was in my twenties, my friends and I were motivated by the eternal frustration of young people—that they are not immediately all made rulers of the world. It seemed supremely annoying to my musician friends, for instance, that the biggest stars—like Michael Jackson—would get millions of dollars in advance for an album, while an obscure minor artist like me would get only a $100,000 advance (and this was in early 1990s dollars).

So what to do? Kill the whole damned system! Make music free to share, and demand that everyone build reputations on a genuine all-to-all network instead of a broadcast network, so that it would be fair. Then we'd all go out and perform to make money, and the best musician would win.

The lecture circuit was particularly good to me as a live performer. My lecture career was probably one of the first of its kind driven mostly by my online presence. (In the old days, my crappy Website got enough traffic to merit coverage by mainstream media such as the *New York Times*.) Money seemed available on tap.

A sweet way to run a culture back then, but in the bigger picture it's been a disaster. Only a tiny, token number of musicians (if any) do as well within the new online utopia as even I used to do in the old world, and I wasn't particularly successful. All the musicians I have been able to communicate with about their true situation, including a lot of extremely famous ones, have suffered after the vandalism of my generation, and the reason isn't abstract but biological.

What we denied was that we were human and mortal, that we might someday have wanted children, even though it seemed inconceivable at the time. In the human species, neoteny, the extremely slow fading of juvenile characteristics, has made child rearing into a draining, long-term commitment.

That is the reality. We were all pissed at our parents for not coming through in some way or other, but evolution has extended the demands of human parenting to the point that it is impossible for parents to come through well enough, ever. Every child must be disappointed to some degree because of neoteny, but economic and social systems can be designed to minimize the frustration. Unfortunately the Internet, as it has come to be, maximizes it.

The way that neoteny relates to the degradation of the Internet is that as a parent, you really can't run around playing live gigs all the time. The only way for a creative person to live with what we can call dignity is to have some system of intellectual property to provide sustenance while you're out of your mind with fatigue after a rough night with a sick kid.

Or spouses might be called on to give up their own aspirations for a career—but there was this other movement, called feminism, happening at the same time, which made that arrangement less common.

Or there might be a greater degree of socialism, to buffer biological challenges, but there was an intense libertarian tilt coincident with the rise of the Internet in the United States. All the options have been ruled out, and the result is a disjunction between true adulthood and the creative life.

The Internet, in its current fashionable role as an aggregator of people through social networking software, values humans only in real time and in a specific physical place—that is, usually away from their children. The human expressions that used to occupy the golden pyramidion of Maslow's pyramid are treated as worthless in themselves.

But dignity is the opposite of real time. Dignity means, in part, that you don't have to wonder if you'll successfully sing for your supper for every meal. Dignity ought to be something one can earn. I have focused on parenting here because it is what I am experiencing now, but the principle becomes even more important as people become ill, and then even more as people age. So for these reasons and many others, the current fashionable design of the Internet, dominated by so-called social networking, has an antihuman quality. But very few people I know share my current perspective.

Dignity might also mean being able to resist the near consensus of your peer group.

Wisdom of the Crowd

Keith Devlin

Executive director, H-STAR Institute, Stanford University; author, The Unfinished Game: Pascal, Fermat, and the Seventeenth-Century Letter That Made the World Modern

In this year's *Edge* question, the key phrase is surely "the way you think," and the key word therein is "think."

No one can contribute to an online discussion forum like this without thereby demonstrating that the Internet has changed, and continues to change, the way we work. The Internet also changes the way we make decisions. I now choose my flights on the basis of a lot more information than any one air carrier would like me to have (except perhaps for Southwest, which currently benefits from the Internet decision process), and I select hotels based on reviews by other customers, which I temper by a judgment based (somewhat dubiously, I admit) on their use of language as to whether they are sufficiently "like me" for their views to be relevant to me.

But is that really a change in the way I *think*? I don't think so. In fact, we *Edge* contributors are probably a highly atypical society grouping to answer this question, since we have all been trained over many years to think in certain analytic ways. In particular, we habitually begin by gathering information, questioning that information and our assumptions, looking at (some) alternatives, and basing our conclusions on the evidence before us.

We are also used to having our conclusions held up to public scrutiny by our peers. Which, of course, is why it is rare—though intriguingly (and I think all to the good) not totally impossible—to find trained scientists who believe in biblical creationism or who doubt that global warming is a real and dangerous phenomenon.

When I reflect on how I go about my intellectual work these days, I see that the Internet has changed it dramatically, but what has changed is the execution process (and hence, on some occasions, the conclusions I reach or the way I present them), not the underlying thinking process.

I would hope for humanity's future that the same is true for all my fellow highly trained specialists. The scientific method for reaching conclusions has served us well for many generations, leading to a length and quality of life for most of us that was beyond the imagination of our ancestors. If that way of thinking were to be replaced by a blind "wisdom of the crowd" approach, which the Internet offers, then we are likely in for real trouble. For wisdom of the crowd—like its best-known exemplar, Google search—gives you the mostly best answer most of the time.

As a result of those two *mostlys*, using the wisdom of the crowd without questioning it, though fine for booking flights or selecting hotels, can be potentially dangerous, even when restricted to experts. To give one example, not many decades ago, the wisdom of the crowd among the scientific community told us that plate tectonics was nonsense. Now it is the accepted theory.

The good thing about the analytic method, of course, is that once there was sufficient evidence in support of plate tectonics, the scientific community switched from virtual dismissal to near total acceptance.

That example alone explains why I think it is good that a few well-informed (this condition is important) individuals question both global warming and evolution by natural selection. Our conclusions need to be constantly questioned. I remain open to having my mind changed on either. But to make that change, I require convincing evidence, which is so far totally lacking. In the meantime, I will continue to accept both theories.

The real *Edge* question, for me, is one that is only implied by the question as stated: Does the Internet change the way of thinking for those people born in the Internet age—the so-called digital natives? Only time can really answer that.

Living organisms adapt, and the brain is a highly plastic organ, so it strikes me as not impossible that the answer to this modified question may be yes. On the other hand, recent research by my Stanford colleague Cliff Nass (and others) suggests that there are limitations to the degree to which the digital environment can change our thinking.

An even more intriguing question is whether the Internet is leading to society as a whole (at least, those who are on the Net) constituting an emergent global thinking. By most practical definitions of "thinking" I can come up with—distinguishing it from emotions and self-reflective consciousness—the answer seems to be yes. And that development will surely change our future in ways we can only begin to imagine.

Weirdness of the Crowd

Robert Sapolsky

Neuroscientist, Stanford University; author, Monkeyluv: And Other Essays on Our Lives as Animals

I should start by saying that I'm not really one to ask about such things, as I am an extremely unsophisticated user of the Internet. I've never sold anything on eBay, bought anything from Amazon, or posted something on YouTube. I don't have an avatar on Second Life and I've never "met" anyone online. And I've never successfully defrauded the wealthy widow of a Nigerian dictator. So I'm not much of an expert on this.

However, like almost everyone else, I've wasted huge amounts of time wandering around the Internet. As part of my profession, I think a lot about the behavior of primates, including humans, and the behavior manifest in the Internet has subtly changed my thinking. Much has been made of the emergent properties of the Internet. The archetypal example, of course, is Wikipedia.

A few years back, *Nature* commissioned a study showing that when it came to accuracy about hard-core scientific facts, Wikipedia was within hailing distance of the *Encyclopaedia Britannica*. Immensely cool—in just a few years, a self-correcting, bottom-up system of quality, fundamentally independent of authorities-from-on-high, is breathing down the neck of the Mother of all sources of knowledge. The proverbial wisdom of crowds. It strikes me that there may be a very interesting consequence of this. When you have generations growing up with bottom-up emergence as routine, when wisdom-of-the-crowd phenomena tell you more accurately what movies you'll like than can some professional movie critic, people are more likely to realize that life, too, can have emerged, in all its adaptive complexity, without some omnipotent being with a game plan.

As another plus, the Internet has made me think that the downtrodden have a slightly better chance of being heard—the efficacy of the crowd. A small example of that: the recent elections in which candidates ran Internet campaigns. Far more consequential, of course, is the ability of the people to vote online about who should win on *American Idol.* But what I'm beginning to think is possible is that someday an abused populace will rise up and, doing nothing more than sitting at their computers and hacking away, freeze a government and bring down a dictator. Forget a Velvet Revolution. What about an Online Revolution? Mind you, amid that optimism, it's hard not to despair a bit at the idiocy of the crowd, as insane rumors careen about the Internet.

However, what has most changed my thinking is the array of oddities online. By this, I don't mean the fact that, as of this writing, 147 million people have watched "Charlie Bit My Finger—Again!" with another 20 million watching the various remixes. That's small change. I mean the truly strange Websites. Like the ones for people with apotemnophilia, a psychiatric disease whose victims wish to lose a limb.

There's someone who sold online, for $263, a piece of gum that Britney Spears had spit out. A Website for people who like to chew on ice cubes. Websites (yes, plural) for people who are aroused by pictures of large stuffed animals "having sex," and one for people who have been cured of that particular taste by Jesus. An online museum of barf bags from airlines from around the world. A site for people who like to buy garden gnomes and stab them in the head with sharp things, and then post pictures of it. And on and on. Weirdness of the crowd.

As a result of wasting my time over the years surfing the Internet, I've come to better understand how people have a terrible craving to find others like themselves, and the more unconventional the person, the greater the need. I've realized that there can be unforeseen consequences in a material world crammed with the likes of barf bags and garden gnomes. And most of all, the existence of these worlds has led me to appreciate more deeply the staggering variety and richness of the internal lives of humans. So, maybe *not* such a waste of time.

The Synchronization of Minds

Jamshed Bharucha

Professor of psychology, provost, senior vice president, Tufts University

Synchronization of thought and behavior promotes group cohesion—for better or worse. People love to share experiences and emotions. We delight in coordinated activity. We feel the pull of conformity. And we feed off each other. Synchronization creates a sense of group agency, in which the group is greater than the sum of the people in it.

The Internet sparks synchronization across vast populations. Never before in history have people been able to relate to each other on this scale. The discovery of new tools has always changed the way we think. We are social beings, and the Internet is the most powerful social tool with which the human brain has ever worked.

Through the Internet, people with common backgrounds, interests, or problems can find one another, creating new groups with new identities in unprecedented ways. Amorphous groups can become energized, as people who had gone their separate ways reconnect. As with all technologies, this powerful social tool can be used constructively or destructively. Either way, it certainly has changed the way we think about ourselves.

People yearn to be part of a group. Most feel part of multiple groups. Group identity is as important to us as anything else and provides the glue that binds us together. Group affiliation is affirming, exhilarating, motivating. As the Internet develops the bandwidth to communicate seamlessly in real time—with more of the nuance of in-person communication—its binding power will become ever more compelling.

The downside of synchronization on this scale is the risk of herding behavior or virtual mobs. However, the transparency and anonymity of the Internet allows contrary feelings to be expressed, which can balance out the narrowing effect of groupthink.

In the early days of the Internet, few predicted it would plug into our social instincts as it has. Not only has the binding force of the Internet changed the way we think about ourselves and the world, it also has possibly enabled an emergent form of cognition—one that occurs when individual minds are intricately synchronized.

My Judgment Enhancer

Geoffrey Miller

Evolutionary psychologist, University of New Mexico; author, Spent: Sex, Evolution, and Consumer Behavior

The Internet changes every aspect of thinking for the often-online human: perception, categorization, attention, memory, spatial navigation, language, imagination, creativity, problem solving, Theory of Mind, judgment, and decision making. These are the key research areas in cognitive psychology and constitute most of what the human brain does. The Websites of BBC News and the *Economist* extend my perception, becoming my sixth sense for world events. Gmail structures my attention through my responses to incoming messages: delete, respond, or star for response later? Wikipedia is my extended memory. An online calendar changes how I plan my life. Google Maps changes how I navigate through my city and world. Facebook expands my Theory of Mind—allowing me to better understand the beliefs and desires of others.

But for me, the most revolutionary change is in my judgment and decision making—the ways I evaluate and choose among good or bad options. I've learned that I can offload much of my judgment onto the large samples of peer ratings available on the Internet. These, in aggregate, are almost always more accurate than my individual judgment. To decide which Blu-ray DVDs to put in my Netflix queue, I look at the average movie ratings on Netflix, IMDb, and Metacritic. These reflect successively higher levels of expertise among the raters—movie renters on Netflix, film enthusiasts on IMDb, and film critics on Metacritic. Any film with high ratings across all three sites is almost always exciting, beautiful, and thoughtful.

My fallible, quirky, moody judgments are hugely enhanced by checking average peer ratings: book and music ratings on Amazon, used car ratings on Edmunds.com, foreign hotel ratings on TripAdvisor, and citations to scientific papers on Google Scholar. We can finally harness the law of large numbers to improve our decision making: the larger the sample of peer ratings, the more accurate the average. As ratings accumulate, margins of error shrink, confidence intervals get tighter, and estimates improve. Ordinary consumers have access to better product rating data than market researchers could hope to collect.

Online peer ratings empower us to be evidence-based about almost all our decisions. For most goods and services—and, indeed, most domains of life—they offer a kind of informal meta-analysis, an aggregation of data across all the analyses already performed by like-minded consumers. Judgment becomes socially distributed and statistical rather than individual and anecdotal.

Rational-choice economists might argue that sales figures are a better indication than online ratings of real consumer preferences, insofar as people vote with their dollars to reveal their preferences. This ignores the problem of buyer's remorse: Consumers buy many things that they find disappointing. Their postpurchase product ratings mean much more than their prepurchase judgments. Consumer Reports.org data on car owner satisfaction ("Would you buy your car again?") are much more informative than sales figures for new cars. Metacritic ratings of the *Twilight* movies are more informative about quality than first-weekend box office sales. Informed peer ratings are much more useful guides to sensible consumer choices than popularity counts, sales volumes, market share, or brand salience.

You might think that postpurchase ratings would be biased by rationalization ("I bought product X, so it must be good or I'd look like a fool"). No doubt that happens when we talk with friends and neighbors, but the anonymity of most online ratings reduces the embarrassment effect of admitting one's poor judgments and wrong decisions.

Of course, peer ratings of any product can, like votes in an election, be biased by stupidity, ignorance, fashion cycles, mob effects, lobbying, marketing, and vested interests. The average online consumer's IQ is only a little above 100 now, and average education is just a couple of years of college. Runaway popularity can be mistaken for lasting quality. Clever ads, celebrity endorsements, and brand reputations can bias the judgment of even the most independent-minded consumers. Rating sites can be gamed and manipulated by retailers. Nonetheless, online peer ratings remain more useful than any other consumer empowerment movement in the last century.

To use peer ratings effectively, we have to let go of our intellectual and aesthetic pretensions. We have to recognize that some of our consumer judgments served mainly as conspicuous displays of our intelligence, openness, taste, or wealth and are not really the most effective way to choose the best option. We have to learn some humility. My best recent movie viewing experiences have all come from valuing the Metacritic ratings over my own assumptions, prejudices, and prejudgments. In the process, I've acquired a newfound respect for the collective wisdom of our species. This recognition that my own thinking is not so different from, or better than, everyone else's is one of the Internet's great moral lessons. Online peer ratings reinforce egalitarianism, mutual respect, and social capital. Against the hucksterism of marketing and lobbying, they knit humanity together into collective decision-making systems of formidable power and intelligence.

Speed Plus Mobs

Alan Alda

Actor, writer, director; host of The Human Spark, *on PBS*

Telephones make me anxious for some reason—so ever since I've been able to communicate over the Web I've seldom gone near the phone. But something strange has happened. At least once a day I have to stop and think about whether what I've just written can be misinterpreted. In e-mail, there's no instant modulation of the voice that can correct a wrong tone, as there is on the phone, and even though I avoid irony when e-mailing anyone who's not a professional comedian or amateur curmudgeon, I sometimes have to send a second note to un-miff someone. This can be a problem with any written communication, of course, but e-mail, Web postings, and texting all tempt us with speed. And that speed can cost us clarity. This is not so good, because increasingly we communicate quickly, without that modulating voice. I'm even one of those people who will e-mail someone across the room.

In addition, the Internet has connected so many millions of us into anonymous online mobs that the impression that something is true can be created simply by the sheer number of people who repeat it. (In the absence of other information, a crowded restaurant will often get more diners than an empty one, not always because of the quality of the food.)

Speed plus mobs. A scary combination. Together, will they seriously reduce the accuracy of information and our thoughtfulness in using it?

Somehow, we need what taking our time used to give us: thinking before we talk and questioning before we believe.

I wonder—is there an algorithm perking somewhere in someone's head right now that can act as a check against this growing hastiness and mobbiness? I hope so. If not, I may have to start answering the phone again.

Repetition, Availability, and Truth

Daniel Haun

Director, the Research Group for Comparative Cognitive Anthropology, Max Planck Institute for Evolutionary Anthropology

I was born in 1977—or 15 B.I., if you like; that is, if you take the 1992 version of the Internet to be the real thing. Anyway, I don't really remember being without it. When I first looked up, emerging out of the dark and quickly forgotten days of a sinister puberty, it was already there. Waiting for me. So it seems to me that it hasn't changed the way I think—not in a before-and-after fashion, anyway. But even if you are reading these lines through graying and uncontrollably long eyebrows, let me reassure you that it hasn't changed the way you think, either. Of course, it has changed the content of your thinking—not just through the formidable availability of information you seek but, most important, through the information you don't. From what little I understand about human thought, though, I don't think the Internet has changed the way you think. Its architecture has not changed yours.

Let me try to give you an example of the way people think. The way you think. I have already told you three times that the Internet hasn't changed the way you think (four and counting), and each time you read it, the statement becomes more believable to you. For more than sixty years, psychologists have been reporting the human tendency to mistake repetition for truth. This is called the illusion-of-truth effect. You believe to be true what you hear often. The same applies to whatever comes to mind first or most easily.

People, including you, believe the examples they can think of right away to be most representative and therefore indicative of the truth. This is called the availability heuristic. Let me give you a famous example. In English, what is the relative proportion of words that start with the letter *K* versus words that have the letter *K* in third position? The reason most people believe the former to be more common than the latter is that they can easily remember a lot of words that start with a *K* but few that have a *K* in the third position. The truth in fact is that there are three times more words with *K* in third than in first position. Now, if you doubt that people really believe this (maybe because *you* don't), you have just proved my point. Availability creates the illusion of truth. Repetition creates the illusion of truth. I would repeat that, but you get my point.

Let's reconsider the Internet. How do you find the truth on the Internet? You use a search engine. Search engines evidently have very complicated ways to determine which pages will be most relevant to your personal quest for the truth. But in a nutshell, a page's relevance is determined by how many other relevant pages link to it. Repetition, not truth. Your search engine will then present a set of ranked pages to you, determining availability. Repetition determines availability, and both together create the illusion of truth. Hence, the Internet does just what you would do. It isn't changing the structure of your thinking, because it resembles it. It isn't changing the structure of your thinking, because it resembles it.

The Armed Truce

Irene M. Pepperberg

Research associate and lecturer, Harvard University; adjunct associate professor, Department of Psychology, Brandeis; author, Alex and Me

The Internet hasn't changed the way I think. I still use the same scientific training that was drummed into me as an undergraduate and graduate student in theoretical chemistry, even when it comes to evaluating aspects of my daily life: Based on a certain preliminary amount of information, I develop a hypothesis and try to refine it so that it differs from any competing equally plausible hypotheses; I test the hypothesis; if it is proved true, I rest my case within the limits of that hypothesis, aware that I may have solved only one piece of a puzzle; if it is proved false, I revise and repeat the procedure.

Rather, what has changed, and is still changing, is my relationship with the Internet—from unabashed infatuation to disillusionment to a kind of armed truce. And no, I'm not sidestepping the question, because until the Internet actually rewires my brain, it won't change my processing abilities. Of course, such rewiring may be in the offing, and quite possibly sooner than we expect, but that's not yet the case.

So, my changing, love-hate relationship with the Internet:

First came the honeymoon phase—believing that nothing in the world could ever be as wondrous, and appreciating the incredible richness and simplicity the Internet brought into my life. No longer did I have to trudge through winter's snow or summer's heat to a library at the other end of campus—or even come to campus—to acquire information or connect with friends and colleagues all over the world.

Did I need to set up a symposium for an international congress? Just a few e-mails, and all was complete. Did I need an obscure reference or that last bit of data for the next day's PowerPoint presentation while in an airport lounge, whether in Berlin or Beijing, Sydney or Salzburg? Ditto. Did I need a colleague's input on a tricky problem, or to provide the same service myself? Ditto. Even when it came to needing to rapidly research and send a gift because I forgot a birthday or anniversary—ditto. A close friend moves to Australia? No problem staying in touch anymore. But here the only things that changed were the various limitations on the types of information accessible to me within certain logistical boundaries.

Next came the disenchantment phase—the realization that more and faster were not always better. My relationship with the Internet began to feel oppressive, overly demanding of my time and energy. Just because I can be available and can work 24/7/365, must I? The time saved and the efficiencies achieved began to backfire. I no longer had the luxury of recharging my brain by observing nature during that walk to the library, or by reading a novel in that airport lounge.

E-mails that supplanted telephone calls were sometimes misunderstood, because vocal modulations were missing. The number of requests to do X, Y, or Z began to increase exponentially, because (for example) it was far easier for the requesters to shoot me a question than to spend the time digging up the answers themselves, even on the Internet. The lit search I performed on the supposedly infinite database failed to bring up that reference I needed—and knew existed, because I read it a decade ago but didn't save it, because I figured I could always bring it up again.

This Internet relationship was supposed to enable all my needs to be met. How did it instead become the source of endless demands? How did it end up draining away so much time and energy? The Internet seemed to have given me a case of attention deficit disorder. Did it really change the way I think or just make it more difficult to have the time to think? Most likely the latter, because judicious use of the off button allowed a return to normalcy.

Which brings me to the armed truce—an attempt to appreciate the positives and accept the negatives, to set personal boundaries and refuse to let them be breached. Of course, maybe it is just this dogmatic approach that prevents the Internet from changing the way I think.

More Efficient, but to What End?

Emanuel Derman

Professor of financial engineering, Columbia University; principal, Prisma Capital Partners; former head, Quantitative Strategies Group, Equities Division, Goldman Sachs and Co.; author, My Life as a Quant: Reflections on Physics and Finance

An engineer, a physicist, and a computer scientist go for a drive. Near the crest of a hill, the engine sputters and stops running.

"It must be the carburetor," says the engineer, opening his toolbox. "Let me see if I can find the problem."

"If we can just push it to the top of the hill, gravity will let us coast down to a garage," says the physicist.

"Wait a second," says the computer scientist. "Let's all get out of the car, shut the doors, open them again, get in, turn the key in the ignition, and see what happens."

I like programming, and when I do it, I'm often unable to stop, because there is always one more easy thing you can try before you get up, one more bug you can try to fix, one more attempt you can make to find the cause of a problem, one more shot at incrementally improving something. Because of the interactivity of programming—edit, compile, run, examine, repeat—you can always take a quick preliminary whack at something and see if it works. You can try a solution without understanding the problem completely.

If, as I do, you spend most of your day in front of a computer, then the Internet brings this endless micro-interactivity into your life by providing you with a willing co-respondent. It abhors a vacuum. It can fill up all your available time by breaking it up into smaller and smaller chunks. If you have a moment, you can reply to an e-mail, check Wikipedia, look at the weather, scan your horoscope, read a movie review, watch a video, suffer through an ad. All hurriedly.

One unmitigatedly good thing is the associative memory this facilitates. If you can't remember the name of the abstract expressionist you read about in an article fifteen years ago in the *New York Times*, an artist who used to live on Old Slip in New York in the 1950s with his then-wife, a French actress who (you recall) was in *Last Year at Marienbad*, you can go to IMDb, look up the movie, find her name, look her up on Wikipedia, and discover that her husband was Jack Youngerman. When I do this a second time now, for verification, I go off on a tangent and discover that she acted with Allen Ginsberg in *Pull My Daisy*. And that she is buried in Cimetière du Montparnasse, one of the more restful places to be buried, not far from where Hemingway used to drink and write at the . . .

But I digress.

Some people say the Internet has made us more efficient. I waste many hours each day being efficient. Efficiency should be a means, not an end. The big question, as always, is: How shall I live?

The Internet hasn't changed the way I think about that. What's changed the way I think about big things (as always) are the people I talk to and the books I read.

I Have Outsourced My Memory

Charles Seife

Professor of journalism, New York University; former journalist, Science; *author,* Proofiness: The Dark Arts of Mathematical Deception

The process was so gradual, so natural, that I didn't notice it at first. In retrospect, it was happening to me long before the advent of the Internet. The earliest symptoms still mar the books in my library. Every dog-eared page represents a hole in my memory. Instead of trying to memorize a passage in the book or remember an important statistic, I took an easier path, storing the location of the desirable memory instead of the memory itself. Every dog-ear is a meta-memory, a pointer to an idea I wanted to retain but was too lazy to memorize.

The Internet turned an occasional habit into my primary way of storing knowledge. As the Web grew, my browsers began to bloat with bookmarked Websites. And as search engines matured, I stopped bothering even with bookmarks; I soon relied on AltaVista, HotBot, and then Google to help me find—and recall—ideas. My meta-memories, my pointers to ideas, started being replaced by meta-meta-memories, by pointers to pointers to data. Each day, my brain fills with these quasi-memories, with pointers, and with pointers to pointers, each one a dusty IOU sitting where a fact or idea should reside.

Now when I expend the effort to squirrel memories away, I store them in the clutter of my hard drive as much as in the labyrinth of my brain. As a result, I spend as much time organizing them and making sure I can retrieve them on demand as I do collecting them. My memories are filed in folders within folders within folders, easily accessible—and searchable, in case my meta-memory of their location fails. And when a file becomes corrupt, all I am left with is a pointer, a void where an idea should be, the ghost of a departed thought.

The New Balance: More Processing, Less Memorization

Fiery Cushman

Postdoctoral fellow, Mind/Brain/Behavior Interfaculty Initiative, Harvard University

The Internet changes the way I behave, and possibly the way I think, by reducing the processing costs of information retrieval. I focus more on knowing how to obtain and use information online and less on memorizing it.

This tradeoff between processing and memory reminds me of one of my father's favorite stories, perhaps apocryphal, about studying the periodic table of the elements in his high school chemistry class. On their test, the students were given a blank table and asked to fill in names and atomic weights. All the students agonized over this assignment, except for one. He simply wrote, "The periodic table can be found inside the back cover of our textbook, including the full name and atomic weight of each element."

What the smart-aleck ninth grader probably didn't realize was that he manipulated one of the most basic trade-offs that governs the performance of brains, computers, and other computational systems. The teacher reckoned that the most efficient way to solve chemistry problems was a memory-intensive solution, holding facts about elements in a brain. The student reckoned that it was more efficient to solve chemistry problems with a process-intensive solution, retrieving facts about elements from books.

In a world where chemistry books are hard to obtain (i.e., processing is expensive), the teacher has the right solution. In a world where chemistry books are easy to obtain (i.e., processing is cheap), the student has the right solution. A few decades ago, you would walk to the library for encyclopedias, books, and maps. Today I access them from my pocket. This fact is easy to recite, but it's important to emphasize just how different the costs of processing are in these two cases. Suppose it takes about twenty minutes to walk to the library and about five seconds to pull out an iPhone and open up the Web browser. The processing demands on me are 1/240 as great as they were for my father. By analogy, my computer has a 2.4 gigahertz processor. A processor 1/240 as powerful operates at 10 megahertz—just a touch faster than the original Macintosh, released in 1984. Computers today operate very differently because of their vastly increased processing power. It would be surprising if I didn't, too.

How has the Internet changed my behavior? When I walk out the door with my suitcase, I usually don't know what airline I'm flying on, what hotel I'll be staying in, how to get to it, where or when my first meeting will be, where a nearby restaurant is for dinner, and so on. A few years ago, I would have spent a few moments committing those details to memory. Now, I spend a few moments finding the "app for that."

After I see a good talk, I forget many of the details—but I remember to e-mail the author for the slides. When I find a good bottle of wine, I take a picture of the label. I don't have to skim an interesting-looking paper as thoroughly before I file it, as long as I plug a few good keywords into my reference manager. I look up recipes after I arrive at the supermarket. And when a friend cooks a good meal, I'm more interested to learn what Website it came from than how it was spiced. I don't know most of the American Psychological Association rules for style and citation, but my computer does. For any particular "computation" I perform, I don't need the same depth of knowledge, because I have access to profoundly more efficient processes of information retrieval.

So the Internet clearly changes the way I behave. It must be changing the way I think at some level, insofar as my behavior is a product of my thoughts. It probably is not changing the basic kinds of mental processes I can perform but it might be changing their relative weighting. We psychologists love to impress undergraduates with the fact that taxi drivers have unusually large hippocampi. But today's taxi drivers have GPS systems. This makes it relatively less important for drivers to memorize locations, and relatively more important for them to quickly read maps. It is a reasonable guess that GPS changes the way that taxi drivers' brains weight memory versus processing; it seems like a reasonable guess that the Internet changes the way my brain does, too.

Often the transformational role of the Internet is described in terms of memory—that is, in terms of the information the Internet stores. It's easy to be awed by the sheer magnitude of data available on Wikipedia, Google Earth, or Project Gutenberg. But what makes these Websites transformative for me is not the data. Encyclopedias, maps, and books all existed long before their titles were dressed up in dots and slashes. What makes them transformative is their availability—the new processes by which that information can be accessed.

The Enemy of Insight?

Anthony Aguirre

Associate professor of physics, University of California, Santa Cruz

Recently I wanted to learn about twelfth-century China—not a deep or scholarly understanding, just enough to add a bit of not-wrong color to something I was writing. Wikipedia was perfect! More regularly, my astrophysics and cosmology endeavors bring me to databases such as arXiv, ADS (Astrophysics Data System), and SPIRES (Stanford Physics Information Retrieval System), which give instant and organized access to all the articles and information I might need to research and write.

Between such uses and an appreciable fraction of my time spent processing e-mails, I, like most of my colleagues, spend a lot of time connected to the Internet. It is a central tool in my research life. Yet what I do that is most valuable—to me, at least—is the occasional generation of genuine creative insights. And looking at some of those insights, I realize that essentially none of them has happened in connection with the Internet.

Given the quantity of information and understanding I imbibe online, this seems strange and—because the Internet is omnipresent—worrisome. Insight is surely like happiness and money: You'll get a certain amount through a combination of hope, luck, and effort. But maximizing it requires the more deliberate approach of paying careful attention to whatever increases or decreases it and making judicious decisions on that basis.

In this spirit, I undertook a short exercise. Looking back, I identified ten ideas or insights important to me and for which I could remember the context in which they arose. According to my tally, two were during conversation, one while listening to a talk, one while walking, two while sitting at a desk researching and thinking, and four while writing. None occurred while I was browsing the Web, reading online articles, e-mailing, et cetera. This raises two obvious questions: Why does the Web seem to be the enemy of insight, and what, if anything, should I do about it?

After examining my list, several possibilities come to mind in answer to the first question. One is that information input from the Internet is simply too fast, leaving little mental space or time to process that information, fit it into existing schema, and think through the implications. This is not a fault of the Internet per se. But the Internet, by dint of its sheer volume of information (generally short treatments of individual topics and powerful search capabilities), strongly encourages overly rapid information inhalation. Most talks or lectures, in contrast, have the dubious virtues of being wildly inefficient as information transmitters and containing chunks either boring or unintelligible enough to give one's mind some space to think.

A second possible problem is that, in general, communication with the Web is just about as one-way as reading a book. My insight tally clearly favors active, laborious construction of a train of thought or argument.

A third possibility relates to the type of thinking the Internet encourages. The ability to instantly access information is wonderful for spinning a web of interconnections among ideas and pieces of data. Yet for deep understanding—in particular, the sort that arises from the careful following of one thread of thought—the Internet is not very helpful. I often find the Web's role is more to tempt me from the path and off to the side than to aid in the journey.

Finally, but perhaps most crucially, my experience is that real, creative insights or breakthroughs require prolonged and concentrated time in the wilderness. There are lots of things I don't know, and I get excited when I uncover something I don't know. I've come to think it's important to cultivate a "don't know" mind—one that perceives an interesting enigma and is willing to dwell in that perplexity and confusion. A sense of playful delight in that confusion and a willingness to make mistakes—many mistakes—while floundering about is a key part of what makes insight possible for me. And the Internet? The Internet does not like this sort of mind. The Internet wants us to know, and it wants us to know *right now*; its essential structure is to produce knowing on demand. I worry not only that the Internet goads us to trade understanding for information but also that it makes us too accustomed to instant informational gratification. Its bright light deprives us of spending time in the fertile mystery of the dark.

Others might, of course, have different experiences of the causes and conditions of insight—and also of the Internet. But I'd bet that mine are not uncommon. So, what should be done? A first reaction—to largely banish the Internet from my intellectual life—is both difficult (like most, I'm at least a low-level addict) and counterproductive: Information is important, and the Internet is a unsurpassable tool for discovering and assembling it.

But the exercise suggests to me that this tool should be used in its rightful place and time and with a bit more separation from the creative acts of thinking, deeply conversing, working through ideas, and writing. Perhaps we should think of the Internet not as an extra part of our brain but as a library—somewhere we occasionally go to gather raw materials that we can take away somewhere else, where we have time and space to be bored, to be forced into nondistraction, and to be bewildered, so that we can create an opportunity for something really interesting to happen.

The Joy of Just-Enoughness

Judith Rich Harris

Independent investigator and theoretician; author, No Two Alike: Human Nature and Human Individuality

The Internet dispenses information the way a ketchup bottle dispenses ketchup. At first there was too little; now there is too much.

In between, there was a halcyon interval of just-enoughness. For me, it lasted about ten years.

They were the best years of my life.

The Rise of Internet Prosthetic Brains and Soliton Personhood

Clifford Pickover

Writer; associate editor, Computers and Graphics; *editorial board,* Odyssey, Leonardo, *and* YLEM; *author,* Archimedes to Hawking

With increasing frequency, people around the globe seek advice and social support from other individuals connected via the Internet. Our minds arise not only from our own brains but also from Internet prosthetic brains (IPBs)—those clusters of people with whom we share information and advice through electronic networks. The simple notion of "you" and "me" is changing. For example, I rely on others to help me reason beyond the limits of my own intuition and abilities. Many of my decisions in life are shaped by my IPBs around the globe, who provide advice on a wide range of topics: software, computer problems, health issues, emotional concerns. Thus, when a decision is made, who is the me making it?

The IPBs generated by social network connectivity can be more important than the communities dependent on geographic locality. Through the IPBs, we exchange parts of minds with one another. By the information we post on the Web and the interactions we have, we become IPBs for others. In some ways, when we die, a part of us survives as an IPB in the memories and thoughts of others and also as trails we've left on the Internet. Individuals who participate in social groups, blogs, and Twitter and deposit their writings on the Web leave behind particles of themselves. Before the Internet, most of us rarely left marks on the world, except on our immediate family or a few friends. Before the Internet, even your family knew nothing of you outside four generations. In the "old days," your great-grandchildren might have carried some vestigial memory of you, but it faded like an ember when they died—often you were extinguished and forgotten. I know nothing about my great-grandparents.

However, in the Internet Age, the complete extinguishing never really happens, especially for prominent or prolific users. For example, the number of Internet searchers for something you wrote may asymptotically approach zero over the decades, but it will never quite reach zero. Given the ubiquity of the Internet, its databases, and search engines, someone a hundred years from now may smile at something you wrote or wonder about who you were. You may become part of this future person's own IPB as he navigates through life. In the future, simulacra of you, derived in part from your Internet activities, will be able to converse with future generations.

Moreover, studies show that individuals within your social network have a profound influence on your personal health and happiness through your contacts on the Internet (whom you usually know) and their friends (whom you may not know). Habits and ideas spread like a virus through a vast Web of interconnectivity. Behaviors can sometimes skip links—spreading to a friend of a friend without affecting the person who connects them. In summary, in the age of the Internet, the concept of you and personhood is more diffuse than ever before.

Because your interests, decision-making capabilities, habits, and even health are so intertwined with those of others, your personhood is better defined as a pseudo-personhood that is composed of yourself and the assembly of your IPBs out to at least three degrees of network separation. When we die, the Web of interconnectivity becomes torn, but one's pseudo-personhood, in some sense, continues to spread, like a soliton wave on a shoreless sea of Internet connections.

When Marc Chagall was asked to explain why he became a painter, he said that a painting was like a window through which he "could have taken flight toward another world." Chagall explored the boundaries between the real and unreal. "Our whole inner world is reality," he once wrote, "perhaps more real still than the apparent world."

As the notion of IPBs and soliton personhood expands, this kind of boundary will become even more blurred. The IPBs become of Chagallian importance and encourage the use of new windows on the world. They foster a different kind of immortality, form of being, and flight.

Immortality

Juan Enriquez

CEO, Biotechonomy; founding director, Harvard Business School's Life Sciences Project; author, The Untied States of America: Polarization, Fracturing, and Our Future

The most important impact on my life and yours is that the Internet grants immortality. Think of your old archaeology/sociology/history course, or your visits to various museums. Think of how painstakingly arrowheads, outhouses, bones, beads, textiles, and sentence fragments have been discovered, uncovered, studied, and preserved.

But these few scraps have provided real knowledge while leaving large lagoons filled with conjecture, theories, speculation, and outright fairy tales. Despite this, we still know an awful lot about a very few. Because most of our knowledge of the past depends on very little about very few, the story of very few lives survives.

As we got better at transmitting and preserving data, we learned quite a bit more about much more. Biographies could rely not just on letters, songs, and folktales but also on increasingly complete business ledgers, bills of sale, newspapers, and government and religious records.

By the time of the last great typhoid epidemics and fires in the United States and Europe, we could trace the history of specific houses, families, wells, cows, and outhouses. We could build a specific history of a neighborhood, a family, an individual. But there were still those large lagoons in our knowledge. Not so today. Any electronic archaeologist, sociologist, or historian examining our e-lives would be able to understand, map, compute, contrast, and judge our lives in a degree of detail incomprehensible to any previous generation. Think of a single day of our lives. Almost the first thing we do after turning off the alarm—before brushing our teeth, having our coffee, seeing to a child, or opening a newspaper—is reach for that iPhone or BlackBerry. As it comes on and speaks to us, or we speak through it, it continues to create a map of almost everything in our lives.

Future sociologists and archaeologists will have access to excruciatingly detailed pictures, on an individual basis, of what arrived, what was read, ignored, deleted, forwarded, and responded to. Complement this stream of data with Facebook, Twitter, Google, blogs, newspapers, analyst reports, and Flickr, and you get a far more concrete and complete picture of each and every one of us than even the most extraordinary detail historians have unearthed on the most studied and respected (or reviled) of world leaders.

And by the way, this cache is decentralized. It exists and multiplies at various sites. Digging through the Egyptian pyramids will look like child's play compared to what future scholars will find at Google, Microsoft, the National Security Agency, credit bureaus, or any host of parallel universes.

It is virtually impossible to edit or eliminate most traces of our lives today, and for better or worse we have now achieved what the most powerful Egyptians and Greeks always sought—immortality.

So, how has this newfound immortality affected my thinking? Well, those of a certain age learned long ago, from the triumphs and tragedies of the Greek gods, that there are clear rules separating the mortal and immortal. Trespasses that are tolerated and forgiven in the fallible human being have drastic consequences for gods. In the immortal world, all is *not* forgiven and mostly forgotten after you shuffle off to Heaven.

A Third Replicator

Susan Blackmore

Psychologist; author, Consciousness: An Introduction

The way "I" think? I'm not sure that I know anymore who or what is doing the thinking. That's the question the Internet is forcing me to ask.

When I was just a human being, writing books and research papers or appearing on radio and television, I could happily imagine that "I" wrote my books. I didn't need to question who or what was doing the thinking or having the new ideas. In those days, body, brain, and knowledge were all bound up together in one place. To use an old metaphor, hardware, software, and data were all bound up in one entity; it was reasonable to call it "me."

The Internet has changed all that. It has changed both the nature of selves and the nature of thinking. "I" am no longer just the imagined inner conscious self who inhabits this body but also the smiling face on my Website and the fictional character other people write about in cyberspace. If someone asks, "Who is Sue Blackmore?" this body will have less say in the answer than the questioner's search engine.

The change to thinking itself began gradually. Humans have long outsourced their knowledge to paper and books, so in the old days I would sit at my desk with my typewriter and look up things I needed to know in books in my own or the university's library. Then I got a word processor. This new hardware shifted a little of the work, but all the creative thinking still went on inside my head, taking in countless old memes and bringing them together to make new ones, selecting among the results, and writing just a few of them down.

Then came the Internet. This meant I could communicate with more people, which meant more mixing of ideas but did not change the process fundamentally. The real change was the advent of the World Wide Web. Suddenly—and in retrospect it really does seem to have been sudden—masses of information were available, right there on my desk. Almost overnight, I stopped using the university library. Indeed, I haven't physically been there for years now.

The Web needed search engines, and these changed the world amazingly quickly. By sifting through mountains of data and coming up with relevant items, they took over a large part of what used to be human thinking.

I like to see all this in evolutionary terms. The creativity of an evolutionary process depends on the three processes of copying, varying, and selecting information. First we had genes—replicators that banded together to create organisms. Then we had memes—replicators that worked together to create human minds. Now we have a third replicator and a new process of creative evolution. All those computers, programs, servers, cables, and other essentials of the Internet might once have seemed to be hardly more than an extension of books, typewriters, and telephones, but we should no longer see them that way.

Books, typewriters, and telephones store information or pass it on, but they do not select the information they copy. They can vary the information by poor-quality copying, but they cannot put together old memes to make new ones. The Internet, or parts of the Internet, can.

Out there in cyberspace are search engines and kinds of software that copy, vary, and select information, concocting new combinations and passing them around the globe in microseconds, making the results available to us all. This is truly a new evolutionary process—one that deals in ideas, one that creates images and original texts. Thinking has escaped from the human scale.

These days I still sit at my desk, but I am not just a human being thinking and writing down my thoughts. The keyboard I type on is recognizably like my old typewriter, but the process I am engaged in is nothing like it was before. Now as I write, I jump quickly and often to things other people have written. I call up pages of information selected by software I do not understand and incorporate these into the text I am working on. This new text may go straight onto my Website or a blog, and from there it may, or may not, be picked up by other sites and copied again. Even books partake of this extraordinary creative process, with Google scanning and propagating pages to students, other writers, and bloggers. No one can possibly know where all the copies and fragments of copies have gone, how many times they have been copied or by what process they were selected. Ever more of the copying, varying, and selecting goes on outside human brains and outside human control.

Is the Internet itself thinking? I would say yes—or if not, it is on the verge of doing so. The digital information it passes around is a third replicator, a kind of information that is copied, varied, and selected by the massive machinery of the Internet and the Web.

So how has the Internet changed the way I think? The words I am writing now are far less "mine" than they were before. Indeed, they have been created as much by John Brockman, the *Edge* community, and the entire Internet as by little me. I did not so much write them as they used me to get themselves written.

So the answer is not that the Internet is changing the way I think; it is changing the nature of thinking itself.

Bells and Smoke

Christine Finn

Archaeologist, journalist; author, Artifacts: An Archaeologist's Year in Silicon Valley

I saw in the new decade wrapped against the English Channel chill under one of the few surviving time ball towers in the world. It was hardly a Times Square ball drop but my personal nod to a piece of eighteenth-century tech that was a part of communications history—ergo, a link to the Internet. For years, this slim landmark signaled navigators off the white cliffs of Dover to set their chronometers to Greenwich Mean Time. It was a Twitter ball, with just one message to relay.

History is my way in, this year. I am answering this year's *Edge* question against the deadline, as the answer slips as defiantly as time. The Internet has not only changed the way I think but prompted me to think about those changes over time, weighted by the unevenness of technology take-up and accessibility to the Net.

I encountered the Web as a researcher at Oxford in the mid-1990s. I learned later that I was at Tim Berners-Lee's former college, but I was pretty blasé about being easily online. I saw the Internet more as a resource for messaging, a faster route than the bike-delivered pigeon post. I didn't see it as a tool for digging, and I remained resolutely buried in books. But when I visited nonacademic friends and asked if I could check e-mails on their dial-ups, I began to equate the Net with privilege, via phone bill anxiety. As they hovered nervously, I dived in and out again. The Internet was not a joy but a catch-up mechanism. And for a while I couldn't think about it any other way.

In 2000, something happened. I found myself drawn to write a book about Silicon Valley. Moving frequently between the United

Kingdom and America's East and West Coasts, I began to think about the implications of the Internet and, moreover, about how being unable to get online was starting to affect me. What was I missing intellectually and culturally by being sometimes out of the game? I began to appreciate a new hunger, for a technology that was still forming. I knew that all that information was out there, and I couldn't realize its potential; sometimes I believed that ignorance was bliss. Traveling around America by bus and train for several months was a revelation. At every stop, I tried to get online, which usually meant I waited in line. I relished my log-on gifts: a precious thirty minutes at the New York Public Library, a whole hour at small towns in the Midwest, a grabbed few minutes in a university department before giving a lecture somewhere.

Then—joy!—luxuriating in the always-on technology at my friends' homes in the Bay Area, where even the kitchens had laptops. But as I made those flights east, the differential was widening. I lost hours trawling the streets of European cities for an Internet café, only to feel that it was merely a brushed kiss from a stranger; there would always be someone else in line. I had had a taste, and I knew that tech was building on tech out there in the ether. I was like some Woody Allen character gazing out of a car window into a train full of revelers. Being barred from the Web felt like a personal blow; I'd lost the key to the library.

In 2004, I moved to Rome just as the Indian Ocean tsunami was showing how the Internet could be mobilized for the common good. I made my first post. I began my own blog, charting Rome's art and culture for Stanford's Metamedia Lab. The pope was declining, and by March 2005 St. Peter's square was mushrooming with satellite dishes. In the Sistine Chapel, God and Adam were connecting on Michelangelo's ceiling; outside, fingers were twitching on laptops and cell phones for one of the Internet's seminal news moments. But I heard the news the old-fashioned way. Walking home with a bag of warm pizza, I heard a sudden churning of bells when it was not the marking of the hour. As I ran with the thousands to St. Peter's, I recall feeling moved by these parallel communications; here, people could still be summoned by bells. A few weeks later, while I was watching wide-screen TV in a Roman café, white smoke rose from the Vatican chimney. The ash drifted over the Vatican's ancient walls, morphing into a messaging cacophony of Italian cell phones and clattering keyboards in heaving Internet cafés.

Dare, Care, and Share

Tor Nørretranders

Science writer; consultant; lecturer; author, The Generous Man: How Helping Others Is the Sexiest Thing You Can Do

The more you give, the more you get. The more you share, the more they care. The more you dare, the more is there for you. Dare, care, and share.

The Internet has become the engine of gift economy and cooperation. The simple insight that there is so much more knowledge, data, and wisdom out there than I can ever attend to in a lifetime shows me that life is not about collecting information into a depot of books, theorems, or rote memories. Life is about sharing with others what you have. Use it, share it, gather it when you need it. There is plenty out there.

In ecology, the waste of one organism is the food of another. Plants produce oxygen as a waste product; animals need it to live. We produce carbon dioxide as waste; plants require it. To live is to be able to share your waste.

Human civilization seems to have forgotten that, through centuries of building and isolating waste depots and exploiting limited resources. Now we are learning that it is all about flows: matter, energy, information, social links. They all flow through us. We share them with one another and all other inhabitants of this planet. The climate problem shows us what happens if we ignore the idea that renewable flows are the real stuff, whereas depots and fortresses are illusions in the long run.

The Internet makes us think in the right way: pass it on, let it go, let it flow. Thinking is renewed. Now we need only to change the way we act.

Getting Close

Stuart Pimm

Doris Duke Professor of Conservation Ecology, Duke University; author, The World According to Pimm: A Scientist Audits the Earth

Once upon a time, we had the same world we do now. We knew little about its problems.

Wise men and women pontificated about their complete worlds—worlds that, for some, stretched only to the limits of their city centers, or sometimes only to the borders of their colleges' grounds. This allowed them clever conceits about what was really important in life, art, science, and the rest of it. Lesser minds would come to pay homage and, let's be honest, use the famous library, since that was the only way of knowing what was known and who knew it. The centers ruled and knew it.

Darkness is falling when I see the light on in the lab and stop by to see who else is working late. There's a conversation going on over Skype. It's totally incomprehensible. Even its sounds are unfamiliar. There's no Rosetta Stone software for the language my two students are learning from their correspondent, who sits in a café in a wretched oiltown on the edge of the rain forest in Ecuador. It's spoken only by a few hundred Indians. All but their children were born as nomads, in a forest that has the luck to be sitting on billions of barrels of oil. (I didn't say "good luck.") In a few months, we'll be in that forest. My students will improve their language skills with the Indian women, helping them prepare *chicha* by chewing manioc, spitting it into the bowl, and chewing another mouthful.

With the Internet, what happens in that forest is exactly as close as anything else I want to understand or communicate (give or take the slow phone line or cell phone reception). When an oil company pushes a road far closer to a reserve than it promised, we'll know about it immediately. When some settlers try to clear forest, we'll know just as quickly if they're killing Indians and if the Indians are killing them. So will everyone else.

The Internet is instant news from remote places, with photos to prove it. What we now think about is suddenly much larger, more frightening, and far more challenging than it once was.

The Internet has vastly more coverage of everything: immediate, future, and past. So when we want to know who has signed which oil exploration leases for which tracts of remote forest, the data are not in Duke's library (or anyone else's), but I can get them online from the Websites of local newspapers. And I can do that in the forest clearing, surrounded by those who futures have been signed away. Knowledge is now everywhere. You can find it from everywhere, too.

The Internet has vastly increased the size of the problem set about humanity's future. Some problems now look really puny. They probably always were.

Who does the thinking has changed, too. When knowledge is everywhere, so are the thinkers.

A Miracle and a Curse

Ed Regis

Science writer; author, What Is Life?: Investigating the Nature of Life in the Age of Synthetic Biology

The Internet is not changing the way I think (nor, as far as I know, the way anyone else thinks). To state the matter somewhat naïvely, I continue to think the same way I always thought: by using my brain and my five (or six) senses and considering the relevant available information. I mean, how else can you think?

What it has changed for me is my use of time. The Internet is simultaneously the world's greatest time saver and the greatest time waster in history. I'm reduced to stating the obvious with regard to time saving: The Web embodies practically the whole of human knowledge and most of it is only a mouse click away. An archive search that in the past might have taken a week, plus thousands of miles of travel, can now be done at blitz speeds in the privacy of your own home or office, et cetera.

The flip side, however, is that the Internet is also the world's greatest time sink. This was explicitly acknowledged as a goal by the pair of twenty-something developers of one of the famous Websites or browsers or search engines, I forget which (it may have been Yahoo!), who once said, "We developed this thing so that you don't have to waste time to start wasting time. Now you can start wasting time right away."

As indeed you can. In the newsprint age, I studiously avoided reading the papers on the dual grounds that (1) the news from day to day is pretty much the same ("Renewed fighting in Bosnia," "Suicide bomber kills X people in Y city"), and (2) in most cases you can do absolutely nothing about it anyway. Besides, it's depressing.

These days, though, while the news content remains the same as before, I am a regular reader of the *New York Times* online, plus of course Google News, plus my local paper. Plus I check the stock market many times daily, along with the weather and Doppler radar, blogs (where I sometimes get into stupid, mind-sapping, time-eating flame wars), the listservs I subscribe to, Miata.net (for any spiffy new Miata products or automotive gossip), my e-mail . . . and this doesn't even half cover the Homeric catalog of Internet ships that I sail on from day to day.

Of course, I don't have to do any of this stuff. No one forces me to. I can blame only myself. Still, the Internet is seductive—which is odd, considering that it's so passive an agency. It doesn't actually *do* anything. It hasn't cured cancer, the common cold, or even hiccups.

The Internet is a miracle and a curse. Mostly a miracle.

"The Plural of Anecdote Is Not Data"

Lisa Randall

Physicist, Harvard University; author, Warped Passages

"The plural of anecdote is not data"—but anecdotes are all I have. We don't yet understand how we think or what it means to change the way we think. Scientists are making inroads and ultimately hope to understand much more. But right now, all I and my fellow contributors can do are make observations and generalize.

We don't even know if the Internet changes the way we read. It certainly changes how we do many aspects of our work. Maybe it ultimately changes how our brains process written information, but we don't yet know. Still, the question of how the Internet changes the way we think is an enormous problem, one that anecdotes might help us understand. So I'll tell a couple (if I can focus long enough to do so).

Someone pointed out to me once that he, like me, never uses a bookmark in a book. I always attributed my negligence to disorganization and laziness—the few times I attempted to use a bookmark I promptly misplaced it. But what I realized after this was pointed out is that not using bookmarks was my choice. It doesn't make sense to find a place in a book that you have begun reading but that is so far from your memory that you don't remember having read it. By not using a bookmark, I was guaranteed to return to the last continuous section of text that had actually made a dent in my brain.

With the Internet, we tend to absorb multiple pieces of information about whatever topic we decide we're interested in. Online, we search. Marvin Minsky recently told me that he prefers reading on an electronic device because he values the search function. And I certainly do, too. In fact, I tend to remember the answers to the pointed questions I ask on the Internet better than the information I pick up reading a long book. But there is also the danger that something valuable about reading in a linear fashion, absorbing information internally and processing it as we go along, is lost with the Internet, or even electronic devices—where it is too easy to cheat by searching.

One aspect of reading a newspaper that I've already lost a lot of is the randomness that that kind of reading offers. Today, when I'm staring at a computer screen and have to click to get to an article, I read only those I know will interest me. When I read print papers—something I do less and less—my eyes are sometimes drawn to an interesting piece (or even an advertisement) that I never would have chosen to look for. Despite its breadth, and the fact that I can be so readily distracted, I still use the Internet in a targeted fashion.

So why don't I stick to print media? The Internet is great for disorganized people like me who don't want to throw something away for fear of losing something valuable they missed. I love knowing that everything is still online and that I can find it. I hate newspapers piling up. I love not having to be in an office to check books. I can make progress at home, on a train, or on a plane (when there is enough room between rows to open my computer). Of course, as a theoretical physicist, I could do that before as well—it just meant carrying a lot more weight.

And I do often take advantage of the Internet's breadth, even if my attention is a little more directed. A friend might send me to a Website. Or I might just need or want to learn about some new topic. The Internet also allows me to be bolder. I can quickly get up to speed on a topic I previously knew nothing about. I can check facts, and I can learn others' points of view on any subject I decide is interesting. I can write about subjects I wouldn't have dared to touch before, since I can quickly find out the context that was previously much more difficult to access.

Which brings me to the title quote, "The plural of anecdote is not data." I thought I should check, on Google, who deserves the attribution. It's not entirely clear, but it might be a pharmacologist named Frank Kotsonis, who was writing about the effects of aspartame. I find this particularly funny, because I stopped consuming aspartame due to my personal anecdotal evidence that it made me focus less well. But I digress.

Here's the truly funny aspect of the quote. The original, from Berkeley political scientist Raymond Wolfinger, was exactly the opposite: "The plural of anecdote is data." I'm guessing this depends on what kind of science you do.

The fact is that the Internet provides a wealth of information. It doesn't yet organize it all or process it or arrange it for scientific conclusions. The Internet allows us (as a group) to believe both facts and their opposites; we'll find supporting evidence for all opinions.

But we can attend talks without being physically present and work with people we've never met in person. We have access to all physics papers as they are churned out, but we still have to figure out which are interesting and process what they say.

I don't know how differently we think because of the Internet. But we certainly work differently and at a different pace. We can learn many anecdotes that aren't yet data.

Though all those distracting e-mails and Websites can make it hard to focus!

Collective Action and the Global Commons

Giulio Boccaletti

Physicist, atmospheric and oceanic scientist; associate principal, McKinsey and Company

The Internet has most definitely changed the way I think about collective action and the effect of science on decision making, particularly when it comes to managing the global environment. Three things come to mind: the Internet's role in providing a platform for taking collective action on environmental problems; its ability to focus our collective consciousness from multidisciplinary issues to one problem, the management of planet Earth; and its effect on the pressures that science is subject to as it deals with this new concern for all things planetary.

The global commons in which we operate—water resources; the carbon stock of the atmosphere, land, and oceans; tropical forests—easily exceed national boundaries, making traditional top-down decisions about management difficult. However, these global commons are fully encompassed by global networked information systems, which therefore provide—beyond access to information—a platform that enables the matching of information to action for those who see an opportunity. And if we take a step back from the short-term progress on policy convergence, this is what can be observed across the world:

Businesses and governments are steering productive efforts toward those global commons in what many now call the green economy, using networks to do so. This is a world where farmers in the extensive irrigation systems of the Indus plains of Pakistan or the Australian Murray-Darling basin can find out online, in real time, how much water they are allocated and thus plan their agricultural activities;

where conservation programs for tropical forests in Brazil or Indonesia (critical components of our global strategy to mitigate greenhouse gas emissions) are planned using global mapping technologies; where we can use networked platforms to coordinate millions of individual decisions on consumption and production of energy through smart grids (information-laden networks for power transmission); where weather data can be acted upon across the globe. And where, for the first time, large-scale interventions in Earth's climate, such as attempts to increase carbon capture by the ocean, are being considered by ventures that already assume a fully networked world.

There used to be an edifice of data and theories inaccessible to all except for the few whose job it was to study Earth. In an attempt to create an integrated story, Earth scientists carefully built this edifice with layer after layer of complicated charts—global temperature fields, wind distribution, land use, geology, ice cover—and their theories drew on disparate disciplines to create an ambitious if incomplete picture of what Earth looks like and, most important, how it functions and how it might change.

To the vast majority of the public, though, this endeavor meant little if anything. An International Geophysical Year was proclaimed from July 1957 through December 1958—a sort of race to the Earth—but the only global event to reach the collective consciousness out of this ambitious program was the Soviet launch of Sputnik in October 1957, heralding the beginnings of the race to space instead. When I started in Earth studies more than a decade ago, Earth science departments were struggling to attract the best students away from engineering and physics departments, planetary issues were under the radar for most MBA students, and the closest businesses got to them was using a picture of the globe as their logo. Earth as an integrated collection of large-scale processes was not a consideration for most people; at best, it was an unvarying venue, the place where we lived our lives, a place we could rely on to, well, be there.

The widespread adoption of global information networks changed all this, allowing access to data and theories, often without the mediation of scientists, spreading ideas and encouraging public debate. And that edifice—the colorful maps representing various aspects of the planet's identity, the carefully compiled data—became an interactive, multidimensional space owned by no one and explored by a wide set of agents. It was a space where our planet (and our role in maintaining or exploiting it) became the subject of intense political, economic, and social interest.

The Internet has given rise to one of the largest instances of collective realization witnessed thus far: People, governments, and businesses across the globe have come to understand, more or less at the same time, that the Earth is not an academic abstraction but an entity we interact with, which we can affect by our daily activities and which in turn affects us. The Internet has created what the philosopher Jean-François Lyotard might have thought of as a new concept of knowledge of the Earth—a worldview created not by individuals but by a collective act of negotiation.

Finally, there is the influence of the Internet on science itself and the pressures that science is subject to. As a scientist, I was trained to understand the limits of what the Earth sciences can say—to develop a feel for the inherent uncertainties hidden in the complexity of our planet's observed phenomena. But in what turns out to be a strikingly recursive story, the new conceptual and integrated model for the Earth, born out of the work of thousands of Earth scientists and crystallized in the collective consciousness by global access to information, is having a profound effect on the questions I see science being called to answer.

Now that Earth has been transformed by the Internet from an object of study to an interactive environment that all are able to explore—a place where economic demands and social issues collide with the disciplinary boundaries of the scientific community—science is being forced to confront operational and applied issues. Choices on where to place offshore wind turbines, for example, have a lot to do with where we believe the global circulation of the atmosphere will end up delivering most of the momentum it picks up in the tropics. Concerns about the viability of the hydropower infrastructure are tied to our understanding of variability in the global hydrological cycle. Questions about the future of carbon capture and sequestration are fundamentally tied to our understanding of geology and biogeochemistry.

How should we plan for a changing climate? Where should we invest? What new technologies should we adopt? Such are the questions that science must answer. The challenge lies in making sure that along with the knowledge, the limits of what science can tell us (and therefore the boundaries of what we can do) are not lost in translation as they travel through the Internet.

Informed, Tightfisted, and Synthetic

Laurence C. Smith

Professor of geography and Earth and space sciences, University of California, Los Angeles; author, The World in 2050: Four Forces Shaping Civilization's Northern Future

I remember very well the day when the Internet began changing the way I think. It happened in the spring of 1993, in a drab, windowless computer lab at Cornell. One of my fellow graduate students (a former Microsoft programmer who liked to stay abreast of things) had drawn a crowd around his flickering UNIX box. I shouldered my way in, then became transfixed as his fingers flew over XMosaic, the first widely available Web browser in the world.

XMosaic was only months old. It had been written at the University of Illinois by an undergraduate named Marc Andreessen (a year later he would launch Netscape, its multibillion-dollar successor) and Eric Bina at the National Center for Supercomputer Applications. There were already some Websites up and running. Urged on by his crowd's word search suggestions ("Sex!" "Kurt Cobain!" "Landsat!"), my fellow student lifted the curtain on a new world of commerce, entertainment, and scientific exchange in barely fifteen minutes. A sense that something important was happening filled the lab. By the next day everyone was using XMosaic.

How has my thinking changed since that day in 1993? Like almost everyone, I've become both more addicted to information and more informed. With so much knowledge at my fingertips, I am far less tolerant of my own ignorance. If I don't know something, I look it up. Today I flip through dozens of newspapers a day, when before I barely read one. Too many hours of my life are consumed in this way, and other tasks neglected, but I am perpetually educated in return.

I am now more economics-minded than before. In 1992, if I had to fly someplace, I called the travel agent who worked around the corner and accepted whatever she said was a good fare. Today I thrash competing search engines to shake the last nickel out of a plane ticket. Before shopping online, I hunt and peck for secret discount codes. This superabundance of explicit pricing information has instilled in me an obsessive thriftiness I did not possess before. Doubtless it has helped contribute to thousands of excellent travel agents losing their jobs, and even more hours of time wasted.

The pace and scale of my branch of science have become turbocharged. Unlike before, when scientific data were hard to get, expensive, and prized, my students and I now freely post or download enormous volumes at little or no cost. We ingest streaming torrents of satellite images and climate model simulations in near real time; we post our own torrents online for free use by unseen others around the planet. In a data-rich world, a new aesthetic of sharing, transparency, and collaboration has emerged to supplant the old one of data hoarding and secretiveness. Earth science has become an extraordinarily exciting, vibrant, and fast-advancing field as a result.

Perhaps the most profound change in my thinking is that the new ease of information access has allowed me to synthesize broad new ideas, drawing from fields of scholarship outside my own. It took less than two years for me to finish a book identifying important convergent trends not only in climate science (my area of expertise) but also in globalization, population demographics, energy, political science, geography, and law. While a synthesis of such scope might well have been possible without the light-speed world library of the Internet, I for one never would have attempted it.

Before 1993, my thinking was complacent, spendthrift, and narrow. Now it is informed, tightfisted, and synthetic. I can't wait to see where it goes next.

Massive Collaboration

Andrew Lih

University of Southern California, Annenberg School of Communication and Journalism; author, The Wikipedia Revolution

What has changed my way of thinking is the ability of the Internet to support the deliberative aggregation of information, through filtering and refinement of independent voices, to create unprecedented works of knowledge.

Wikipedia is the greatest creation of massive collaboration so far. That we have a continuously updated, working draft of history capturing the state of human knowledge down to the granularity of each second is unique in human experience.

Wikipedia and now Twitter as generic technical platforms have allowed participants to modify and optimize the virtual workspace to evolve new norms through cultural negotiation. With only basic general directives, participants implicitly evolve new community conventions through online stigmergic collaboration.

With the simple goal of writing an encyclopedia, Wikipedians developed guidelines regarding style, deliberation, and conflict resolution and crafted community software measures to implement them. In the Twitter universe, retweeting and hashtags organically crafted by users extended the microblogging concept to fit emerging community desires. This virtual blacksmithing in both the Wikipedia and Twitter workspaces supports a form of evolvable media "impossibly" supported by the Internet.

So far, our deep experiences with this form of collaboration have been in the domain of textual data. We see this also in journalistic endeavors that seek truth in public documents and records. News organizations such as *Talking Points Memo* and the *Guardian* have mobilized the crowd to successfully tackle hundreds of thousands of pages of typically intractable data dumps. Mature text tools for searching, differential comparison, and relational databases have made all this possible.

We have only started to consider the implications in the visual and multimedia domain. Today we lack the sufficient tools to do so, but we will see more collaborative creation, editing, and filtering of visual content and temporal media. Inevitably the same creative stigmergic effect in the audiovisual domain from Internet-enabled collaboration will result in works of knowledge beyond our current imagination.

It's hard to predict exactly what they will be. But if you had asked me in 2000 whether something like Wikipedia was possible, I would have said, "Absolutely not!"

We Know Less About Thinking Than We Think

Steven R. Quartz

Neuroscientist; associate professor of philosophy, Caltech; coauthor (with Terrence Sejnowski), Liars, Lovers, and Heroes: What the New Brain Science Reveals About How We Become Who We Are

I don't know how the Internet is changing the way I think, because I don't know how I think. For that matter, I don't think we know very much about how anyone thinks. Most likely our current best theories will end up relegated to the dustbin as not only wrong but misleading. Consider, for example, our tendency to reduce human thought to a few distinct processes. We've been doing this for a long time: Plato divided the mind into three parts, as did Freud. Today, many psychologists divide the mind into two (as Plato observed, you need at least two parts to account for mental conflict, as in that between reason and emotion). These dual-systems views distinguish between automatic and unconscious intuitive processes and slower and deliberative cognitive ones. This is appealing, but it suffers from considerable anomalies. Deliberative, reflective cognition has long been the normative standard for complex decision making—the subject of decision theory and microeconomics. Recent evidence, however, suggests that unconscious processes may actually be better at solving complex problems.

Based on a misunderstanding of its capacity, our attention to normative deliberative decision making probably contributed to a lot of bad decision making. As attention turns increasingly to these unconscious, automatic processes, it is unlikely that they can be pigeonholed into a dual-systems view. Theoretical neuroscience offers an alternative model with three distinct systems—a Pavlovian system, a habit system, and a goal-directed system, each capable of behavioral control. Arguably, this provides a better understanding of human decision making: The habit system may guide us to our daily Starbucks fix (even if we no longer like their coffee), whereas the Pavlovian system may cause us to choose a pastry once we're there, despite our goal of losing weight. But this model, too, probably severely underestimates the number of systems constituting thought. If a confederacy of systems constitutes thought, is the number closer to 4 or 400? I don't think we have much basis today for answering one way or another.

Consider also the tendency to treat thought as a logic system. The canonical model of cognitive science views thought as a process involving mental representations, and rules for manipulating those representations (a language of thought). These rules are typically thought of as a logic, which allows various inferences to be made and allows thought to be systematic (i.e., rational).

Despite more than a half century of research on various logics (once constituting the entire field of non-monotonic logics), we still don't know even the broad outlines of such a logic. Even if we did know more about its form, it would not apply to most thought processes. That is, most thought processes appear not to conform to cognitive science's canonical view of thought. Instead, much of thought appears to rest on parallel, associative principles—all those currently categorized as automatic, unconscious ones, including probably most of vision, memory, learning, problem solving, and decision making. Here, neural network research, theoretical neuroscience, and contemporary machine learning provide suggestive early steps regarding these processes but remain rudimentary. The complex dynamics underlying nonpropositional forms of thought are still an essential mystery.

We also know very little about how brain processes underlie thought. We do not understand the principles by which a single neuron integrates signals, nor even the "code" it uses to encode information and signal it to other neurons. We do not yet have the theoretical tools to understand how a billion of these cells interact to create complex thought. How such interactions create our inner mental life and give rise to the phenomenology of our experience (consciousness) is as much a fundamental mystery today as it was centuries ago.

Finally, there is a troubling epistemological problem: In order to know whether the Internet is changing how I think, my introspection into my own thinking would have to be reliable. Too many clever psychology and brain-imaging experiments have made me suspicious of my own introspection. In place of the Cartesian notion that our mind is transparent to introspection, it is very likely that numerous biases undermine the possibility of self-knowledge, making our thinking as impermeable to ourselves as it is to others.

An Impenetrable Machine

Emily Pronin

Associate professor of psychology, Princeton University

A subject in a psychology experiment stands in a room with various objects strewn around the floor and two cords hanging from the ceiling. He is tasked with finding ways to tie the two cords together. The only problem is that they are far enough apart so that if he grabs one, he can't reach the other. After devising some obvious solutions (such as lengthening one of the cords with an extension cord), the subject is stumped. Then the experimenter casually bumps into one of the cords, causing it to swing to and fro. The subject suddenly has a new idea! He weights one of the cords and sets it swinging pendulum-style, then grabs the other cord and waits until the first one swings close enough to catch.

Here's something interesting about this experiment: Most of the subjects failed to recognize the experimenter's role in leading them to this new idea. They believed that the thought of swinging the cord just dawned on them, or resulted from systematic analysis, or from consulting physics principles, or from images they conjured of monkeys swinging in trees. As this experiment and others like it (reviewed in a classic article by Richard Nisbett and Timothy Wilson) illustrate, people are unaware of the particular influences that produce their thoughts.* We know what we think, but we don't know why we think it. When a friend claims that it is her penchant for socialist ideals that leads her to support the latest health care reform bill, it might be wise for you to assume she likes the bill but to doubt her reasons why (and she ought to share your skepticism).

---

* "Telling More Than We Can Know: Verbal Reports on Mental Processes," *Psychological Review* 84, 3 (1977): 231–59.

This brings me to the question of how the Internet has changed the way I think. The problem is this: When it comes to my thoughts, I can honestly tell you what I think (about everything from mint-chip ice cream to e-mail; I love the former and am ambivalent about the latter), but I can only speculate as to why I think those things (does my love of mint-chip ice cream reflect its unique flavor or fond childhood memories of summer vacations with my parents before their divorce?). How has the Internet changed the way I think? I can't really say, because I have no direct knowledge of what influences my thinking.

The idea that my own mental processes are impenetrable to me is a tough one to swallow. It's hard to accept the idea that at a basic level I don't know what's going on in my own head. At the same time, the idea has a certain obviousness to it: Of course I can't recount the enormous complexity of biochemical processes and neural firing that gives rise to my thoughts. The typical neuron in my brain has thousands of synaptic connections to other neurons. Sound familiar?

The Internet's most popular search tool also feeds me thoughts (tangible ideas encoded in words) via a massively connected system that operates in a way that is hidden to me. The obscurity of Google's inner workings (or the Net's, more generally) makes its potential effect on my thoughts somewhat unnerving. My thinking may be influenced by unexpected search hits and extraneous words and images derived via a process beyond my comprehension and control. So although I have the feeling that it's me driving the machine, perhaps it's more the machine driving me. But wait—hasn't that always been the case? Same process, different machine.

A Question Without an Answer

Tony Conrad

Experimental filmmaker; musician/composer

The *Edge* question is a question without an answer, like "When did you stop beating your husband?" It speaks across a divide that's transparent in language but not in social structuring. Even if you decide to disagree with the late Niklas Luhmann, he was the clearest spokesperson on this point when he wrote: "Systems of the mind and systems of communication exist completely independently of each other . . . however, they form a relationship of structural complementarity," and "The independence of each closed system is a requirement for structural complementarity, that is, for the reciprocal initiation (but not determination) of the actualized choice of structure."

What are we trying to deal with here? The question itself—"How is the Internet changing the way you think?"—intimates that the Internet has in some sense materially infected the structure of (inter) subjectivity, that language itself (that is, our common consciousness and unconscious), in the context of global Internet communications, is altered or somehow functions differently. But in the sense that language is our shared thinking, there is no way the Internet might change only the way "you" think or "I" think. This "thinking" is a mediated collectivity that functions meaningfully not as "you" or "I" but as "we." The question can *only* be "How is the Internet changing the way *we* think?"

Talk around it, don't answer; it's that riddle whose answer is not to be answered.

The Internet has sapped my illocutionary force. Even before the Internet, various communication channels represented radically differentiated registers of illocutionary force; for instance, I could write words I wouldn't use carelessly in civil society. The distance and anonymity of the Internet abets radical illocutionary shifts in language resources. This restaging of modes of address at all levels has impinged especially effectively on me as a culture worker who engages in interpersonal confrontations at various ranges of effective distances between me and the subject.

Internet communication is intimate. Internet activity is usually—almost always—individual, a confrontation between the solo subject and the interface with Everything. Sitting alone at my laptop, I'm shrouded in an enveloping trance, a shaped direction; the Internet is intimate, comes in very close, takes up a station nearly inside the sensorium, internal to the subject and somehow exempting the noumenal world, in a way that is quite privileged. Just as a good novel shares its style of thinking, its language and outlook, with me, the Internet also gets internalized to a great degree. Aha! But then, along with this surface—the "neutral" screen surface that carries this or that here or there—comes a fluid of advertising structures. This also happens in magazines and TV, but their surfaces are far less fluid, the distances between information strategies and attentional manipulation far greater, more explicit.

How is the Internet changing the way you (or I) think? What thinking? Is this intended to focus our attention on focused attention? In short, on trance? "Flow"? Focus? Attention? For me, these areas are guided by unconscious processes (which is what advertisers count on): skeins of desires, memory constructs, social links, affiliations, and associations, bundled in carefully culturally regulated assemblages. I become aware that many tools of advertising are formal correlates of classic trance induction rituals, which in turn are closely paralleled by conceptual and formal art strategies. The trance induction procedure that disrupts focused thinking by counting backward could serve as a conceptual or formal art piece. So could the disruptive spellings of advertising—*kool*, *E-Z*, *kleen*, et cetera—that similarly disrupt focused thinking. The intrusion of advertising onto the Internet hasn't been simply additive; it is multiplicative. Advertising communication modulates my attentional systems; that's its aim. As an artist, I become aware that the function of these formal structures and devices is the command and control of my attention.

In the matrix of intersections among globalization, new media, and jurisprudence are unoccupied spaces tricksters should locate and occupy. What kind of social role does this thinking suggest? Twenty years ago, I proposed that media artists should try to break laws that had not yet been written—or, as I put it in a public address at the time, instead of impotent and fitful gestures to service and educate "the community," the need is really (1) to find creative engagements with the law, to set instrumental moral examples for the new home camcorder user, and (2) to invent new crimes.

Not only artists and outsiders but I and everyone else on the Internet have become "criminalized." In a city, anonymity invites both opportunistic and planned criminality, because there are far fewer chances of being recognized than in a village. Online, my identity is a security negotiation, as at airports and banks. But airports and banks are institutions, isolated from home, whereas the Internet is intimate.

The fact that I can shop at home means that I can steal and be stolen from at home, too. The pervasive capacity for deception that the Internet embodies is directly coupled to intimacy—not only the intimacy of circumstantial space, such as the home, but intimacies of the imaginary: sexual desires, secret wishes, possessiveness and power dreams, hunger and fatigue, ill health and the threat of death. These are Big Things; they bulldoze fears of criminality aside and open the gates to criminal thoughts, on one hand, and victimization, on the other. These "thoughts" are not so much coherent plans as mindsets—a general drift or modality of moral outlook—and this sea change of outlook has made for adjustments in my language, expectations, social commerce, vigilance, and levels of depression or exhilaration, in a proliferating tangle of ways.

By delocalizing, rapidly substituting communications for travel (or perhaps creating a convergence between these two systems), the Internet has transposed the global socioeconomic North/South vector into a socially vertical vector, as "the rich get richer and the poor get poorer." These vectors—North/South and rich/poor—may not commonly be perceived as globally equivalent, but neocolonialism does equal the globalization of wage slavery accompanied by huge bonuses among financial managers, with the larger effect of class structure rigidification and the global assurance of corporate hegemony.

The Internet, which has been the vehicle for this 90-degree shift, comprises also the vehicle for stabilization of the new order, since it is the convergence site for every channel of mass communications. But if everyone has free access to information sources, won't these social disparities be resolved through agonistic processes of one sort or another? Or is it instead possible that the opposition movement will merely sustain a quasi-stable dialectical balance? These are Internet "thoughts" that daily bring me pain.

Analytical understanding doesn't resolve conflicts. In fact, the resolution of social disparities is being addressed more trenchantly today by religious fervor than by academic analysis, so the control of belief systems by the Internet, and its evolving strategies for adjusting our minds, represents the balance of future power. Is this at work on me? Likely so. As it happens, belief, conviction, and knowledge are language structures dangling in the language breeze—a breeze inflected, regulated, and abetted by formal processes that attract me: repetition, metaphorical displacement, tradition and ritual, iconic simplification, bait and switch and other psychological tricks ("You can't have any spinach!" "No! I *want* spinach!"), and framing or setting apart.

The Internet's global agenda is counterpoised to the real-world conditions of my geographic localism. *Community* has become a fluidly negotiable term because of the Internet. However it is conceived, though, community online is radically different from community in a geographically local sense. Second Life offers an extreme example of the former; opposed to this is the way local churches in my real world address community housing and schooling issues in my city. However, when these same churches reach out online, or even on TV, their function is diluted to meaninglessness.

The Internet, by distracting me from local matters of immediate and actionable significance, has destructively interfered with my neighborhood agency. Nevertheless, to the degree that all politics is local, the Internet can be used to expand awareness and interactivity within and of a geographically limited function. It cost-effectively supplements direct-mail public relations, leading neighbors to events and connectedness that otherwise would slip past them. All of this is helpful, but it doesn't account for the urgency with which people believe the Internet is a direct route to power for them, when in fact the Internet is so exploitable by power as a control mechanism. Meanwhile I've begun to think of power differently.

In my West/North world, consensus formation has been atomized as a by-product of Net surfing. I can turn anywhere and find confirmation or contestation of almost anything I happen to have in mind. The intimacy of Internet peer group communications (Facebook, etc.) challenges the parents, tribes, churches, communities, workplaces, and schools whose authority formerly dominated the plane of large-scale belief formation and condensation. Meanwhile, in regions where Internet communication is more rare, more regulated, more obviously slanted, and Net surfing is less stochastic than in my West/North, religious conviction is more coherent and is linked to idioms of authority.

A narrow information channel can pretend to be the parents, tribes, churches, communities, workplaces, and schools that for me are washed away by my surfing relativism. But the illusion of individual empowerment that Internet surfing thrusts upon me is simply the backwash of a tidal rise in the technologies of control, effected in two directions: power's structuring of "freedom" of choice and exchange, and power's concomitant harvesting of data with explicit aims to regulate my real-world behavior.

Internet surfing completely absorbs me in the flux and flow of the present moment, in contrast to reading a book, or learning a machine, or studying with a teacher. These enterprises demand sustained, linear thinking. But my students don't think they need to read a whole book to respond to any given challenge; they can simply go to the Internet and a search engine will "think outside the box" for them. This has made me despondent about a general degradation in people's habituation to focused linear thinking.

Today I opened an e-mail message that asked me how the Internet is changing the way I think. I've received thousands of phone calls but never gotten one that asked how the telephone has affected my thinking. I've read many books without ever coming across this question about books, at least put so directly. The same with speaking to friends or watching movies. Now, in general, I regard the rise of inventions such as the Internet from a constructivist perspective; in this instance, as a consequence of the built-out social needs—among capital and the military—for telegraphy, telephones, fax, and so forth. So what is it about the Internet, then? Which social necessity made it so singularly reflexive?

Conceptual Compasses for Deeper Generalists

Paul W. Ewald

Professor of biology, Amherst College; author, Plague Time: How Stealth Infections Cause Cancer, Heart Disease, and Other Deadly Ailments

When I was a kid growing up in Illinois in the early sixties, my mother took me on weekly trips to the Wilmette Public Library. It was a well-stocked warren of interconnected sand-colored brick buildings that grew in increments as Wilmette morphed from farmland to modest houses interspersed with vacant lots, to an upwardly mobile bland Chicago suburb, and finally to a pricey bland Chicago suburb. My most vivid memory of those visits was the central aisle, flanked by thousands of books reflecting glints of "modern" fluorescent lights from their crackly plastic covers. I decided to read them all. I began taking out five books each weekend with the idea that I would exchange them for another five a week later and continue until the mission was accomplished. Fortunately for my adolescence, I soon realized a deflating fact: The Wilmette library was acquiring more than five books per week.

The modern Internet has greatly increased the availability of information, both the valuable stuff and the flotsam. Using a conceptual compass, a generalist can navigate the flotsam to gain the depth of a specialist in many areas. The compass-driven generalist need no longer be dismissed as the Mississippi River—a mile wide and a foot deep.

My current fixation offers an illustration. I'm trying to develop a unified understanding of the causes of cancer. This goal may seem like a pipe dream. Quick reference to the Internet seems to confirm this characterization. Plugging "cancer" into Google, I got 246 million hits, most of them probably flotsam. Plugging "cancer" into PubMed, I got 2.4 million scientific works. Some of these will be flotsam, but most have something of value. If I read ten papers per day every day, I could read all 2.4 million papers in 657 years. These numbers are discouraging, but it gets worse. PubMed tells me that in 2009 there were 280 articles on cancer published per day. Memories of the Wilmette Public Library loom large.

I navigate through this storm of information using my favorite conceptual compass: Darwin's theory of evolution by natural selection. Application of evolutionary principles often draws attention to paradoxes and flaws in arguments. These problems, if recognized, are often swept under the rug, but they become unavoidably conspicuous when the correct alternative argument is formulated. One of my research strategies is to identify medical conventional wisdom that is inconsistent with evolutionary principles. Next I formulate alternative explanations that are consistent, and then I evaluate all of them with evidence.

In the case of cancer, expert opinion has focused on mutations that transform well-behaved cells into rogue cells. This emphasis (bias?) has been so narrow that experts have dismissed other factors as exceptions to the rule. But it raises a paradox: The chance of getting the necessary mutations without destroying the viability of the cell seems much too low to account for the widespread occurrence of cancers. Paramount among the cancer-inducing mutations are those that disrupt regulatory processes that have evolved to prevent damage from cancer and other diseases' cell proliferation. One of these barriers to cancer is the arrest of cellular replication. Another is a cap on the total number of cell divisions. Still another is the tendency for cells to commit suicide when genetic damage is detected.

For a century, research has shown that infections can cause cancer. For most of this time, this knowledge was roundly dismissed as applying only to nonhuman animals. Over the past thirty years, however, the connection between infection and human cancer has become ever stronger. In the 1970s, most cancer experts concluded that infection could be accepted as a cause of no more than 1 percent of human cancer. Today, infectious causes are generally accepted for about 20 percent of human cancer, and there's no end in sight for this trend.

When infections were first found to cause cancer, experts adjusted their perspective by the path of least resistance. They assumed that infections contribute to cancer because they increase mutation rates. An alternative view is that infectious agents evolve to sabotage the barriers to cancer. Why? Because barriers to cancer are also barriers to persistence within a host, particularly for viruses. By causing the cells they live in to divide in a precancerous state, viruses can survive and replicate below the immunological radar.

The depth of biological knowledge and the ability of the Internet to access this depth allows even a generalist to evaluate these two alternative explanations. Every cancer-causing virus that has been well studied is known to sabotage these barriers. Additional mutations (some of them perhaps induced by infection) then finish the transformation to cancer.

Which viruses evolve persistence? This question is of critical practical importance, because we are probably in the midst of determining the full scope of infection-induced cancer. Evolving an ability to commandeer host cells and drive them into a precancerous state is quite a feat, especially for viruses, which tend to have only a dozen or so genes. To evolve mechanisms of persistence, viruses probably need a long time, or very strong selective pressures over a short period of time. Evolutionary considerations suggest that transmission by sex or high-contact kissing could generate such strong selection, because the long intervals between changes in sex or kissing partners (for most people) places a premium on persistence within an individual.

The literature on human cancer viruses confirms this idea—almost all are transmitted by kissing or by sex.

The extent to which this information improves quality and quantity of life will depend on whether people get access to it and alter their behavior to reduce their risk. The earlier the better, because exposure to these viruses rises dramatically soon after puberty. Luckily, kids now have broad access to information before they have access to sexual partners. It will be tougher for the rest of us, who grew up before the modern Internet, in the primitive decades of the twentieth century.

Art Making Going Rural

James Croak

Artist

When the Sui Dynasty sent the literati, the scholar-gentry, to teach Confucian classics to the unschooled in China's farmlands, they dragged carts of calligraphy and paintings to isolated hamlets throughout the vast countryside. For centuries, it was how culture was dispersed in China and among some islands of Japan. Today they would drag a cable.

The Internet allowed me to move to the countryside and make sculpture in the open snowy woods instead of the dark canyons of New York City. I resided in urban centers, especially New York, for most of my adult life, but in my spare time I was drawn to rural places: sojourns to the Gulf of Mexico, sabbaticals to the Rockies, treks into the Arizona desert among the saguaro and devil's claw. Those places never seemed to be suitable for work—too isolated—until the Internet.

I always loved raw nature, but I saw it as antithetical to contributing to the cultural world that centers in a large city. But a gradual thing happened while I was situated in that nexus, Manhattan: The Internet grew up around me. Trips to the library became trips to my screen; art house movies gave way to, well, trips to my screen, where YouTube and Netflix provide a private movie house living on my desk. The daily lift ride to my postal box became several trips to the screen each day, as fountain pens and stamps gave way to instant chatter among friends and not-friends. Taxi rides to supply shops gave way to Internet orders; let UPS lug it home. Negotiating the racks of neighborhood bookstores gave way to browsing Amazon with its reams of attached reviews. The pluriform reasons to live in a metropolis were appearing on my desk and not out past my doorman.

The dawning happened during a photo trip to the Everglades. I took my computer with me—not just the phone or a dim laptop, but the big screen—to a strip motel whose swinging sign bragged "Internet" in perhaps the best Palmer script ever painted in peacock blue. There atop the lauan was the same view that I had in NYC: the *New York Times* Website, an FTP site, rows of e-mail, my bank's Website with a new charge for Conch Shell Fantasy swallowed an hour earlier. Our common nervous system had followed me into the sea of grass, and I knew right then that I would follow that blinking cable farther into the countryside.

Robert Frost wrote about arriving in a place in the woods so deep even his horse was puzzled:

> He gives his harness bells a shake
> To ask if there is some mistake.
> The only other sound's the sweep
> Of easy wind and downy flake.

I did follow that cable into the country, among trees not felled for newspapers not printed because home delivery is the Web. In keeping with this revelation, a Kindle will come next. My shelves cluttered with a lifetime of collected books will not increase, at least not at the previous steady rate; instead they will give way to the electric tablet, as rewritable as the clay tablets of Babylon but with a magic cuneiform from the Wi-Fi spirit that hovers in the air and inscribes stories for me to read. I need only petition this Wi-Fi for a story or a daily newspaper and the reed begins carving within sixty seconds. Rewritable tablet to fixed paper to rewritable tablet in only 6,000 years, and all three can be read under a tree.

Can one telecommute as a sculptor? I send images of new pieces to my dealer; we discuss where they might land. As sculpture is not flat, it cannot be confused with a blunted JPEG as perhaps painting can, a qualitative flattening that worries painters. Larger works can be created simply because I have more space in the country. Living in Manhattan, I thought in terms of square feet; in the country, I think in terms of acres. After e-mail, my most basic Internet task is using the Net as a photo library, often eliminating the need to track down and hire life models. Need an eleven-year-old wearing a long bathing suit, twisting to the left, with hands in the air? Give me two minutes and I will have it, often from multiple angles, printed out and stapled to the wall of my studio. Internet means figurative accuracy.

It also means dialog among like-minded people. I presently have four banters—e-mail threads—under way with art or architecture students in as many countries. Students are not bashful about sending notes out of the blue requesting recipes for making this or that, or advice on education or on how to prevail as an artist, et cetera. I answer most, out of a curiosity about what is on the next generation's mind, hoping to keep my own mind pliable. The barriers preventing the student from reaching out to the experienced have fallen—no longer a letter passed from publisher to dealer to artist over a month or two but instead a note read at breakfast and a response by lunch.

For me, the Internet has made art making rural, not centered in cities as it had been for centuries.

The Cat Is Out of the Bag

Max Tegmark

Physicist, MIT; researcher, Precision Cosmology; scientific director, Foundational Questions Institute

I have a love-hate relationship with the Internet. Maintaining the Zen-like focus so crucial for doing science was easier back when the newspaper and the mail came only once a day. Indeed, as a part of an abstinence-based rehab program, I now try to disconnect completely from the Internet while thinking, closing my mail program and Web browser for hours, much to the chagrin of colleagues and friends, who expect instant response. To get fresh and original ideas, I typically need to go even further and turn off my computer.

On the other hand, the Internet gives me more time for such Internet-free thinking, by eliminating second-millennium-style visits to libraries and stores. The Internet also lets me focus my thinking on the research frontier rather than on reinventing the wheel. Had the Internet existed in 1922 when Alexander Friedmann discovered the expanding-universe model, Georges Lemaître wouldn't have had to rediscover it five years later.

The Internet gives me not only traditionally available information faster (and sometimes faster than I can retrieve it from memory) but also previously unavailable information. With some notable exceptions, I find that "the truth, nothing but the truth, but maybe not the whole truth" is a useful rule of thumb for news reporting, and I usually find it both easy and amusing to piece together what actually happened by pretending I just arrived from Mars and comparing a spectrum of Websites from Fox News to Al Jazeera.

The Internet also affects my thinking by leaving me thinking about the Internet. What will it do to us? On the flip side, as the master of distraction, it seems to be further reducing our collective attention span from the depths to which television brought it. Important issues fade from focus fast, and while many of humanity's challenges get more complicated, society's ability to pay attention to complex arguments dwindles. Sound bites and attack ads work well when the world has attention deficit disorder.

Nevertheless, the ubiquity of information is clearly having a positive effect in areas ranging from science and education to economic development. The essence of science is to think for oneself and question authority. I therefore delight in the fact that the Internet makes it harder to restrict information and block the truth. Once the cat is out of the bag and in the Cloud, that's it. Today it's hard even for Iran and China to prevent information dissemination. Soviet-style restrictions on copying machines sound quaint today, and the only currently reliable censorship is not to allow the Internet at all, as in North Korea.

Love it or hate it, free information will transform the world. Oft-discussed examples range from third-world education to terrorist technology. As another example, suppose someone discovers and posts online a low-tech process for mass-producing synthetic cocaine, THC, or heroin from cheap and readily available chemicals—much like methamphetamine manufacturing today, except safer and cheaper. This would trigger domestic drug production in industrialized countries that no government could stop, in turn slashing prices and potentially devastating the revenue and the power of Colombian and Mexican drug cartels and the Taliban.

Everyone Is an Expert

Roger Schank

Psychologist and computer scientist; founder, Engines for Education, Inc.; author, Making Minds Less Well Educated Than Our Own

The Internet has not changed the way I think, nor has it changed the way anyone else thinks. Thinking has always been the same. To simplify: The thinking process starts with an expectation or hypothesis, and thinking requires one to find (or make up) evidence that explains where that expectation went wrong and to decide upon explanations of one's initial misunderstanding. The process hasn't changed since caveman times. The important questions in this process are these: What constitutes evidence? How do you find it? How do you know if what you found is true? We construct explanations based on the evidence we have found. What has changed is how we find evidence, how we interpret the evidence we have found, and how we find available explanations from which to choose.

I went into AI to deal with exactly this issue. I was irritated that people would argue about what was true. They would get into fights about Babe Ruth's lifetime batting average. That doesn't happen much anymore. Someone can quickly find it. Argument over.

At first glance, we might think the Internet has radically changed the way we look for and accept evidence. I'm sure this is true for the intellectuals who write *Edge* response essays. I'm able to find evidence more quickly, to find explanations that others have offered more easily. I can think about a complex issue with more information and with the help of others who have thought about that issue before. Of course, I could always do this in a university environment, but now I can do it while sitting at home, and I can do it more quickly. This is nice, but less important than people realize.

Throughout human history, evidence to help thinking has been gathered by consulting others, typically the village elder, who might very well have gotten his knowledge by talking to a puff of smoke. Today people make decisions based on evidence they get from the Internet, all right, but that evidence often is no better than the evidence the village elder may have supplied. In fact, that evidence may have been posted by the modern-day version of the village elder.

The intelligentsia may well be getting smarter because they have easy access to a wider range of good thinking, but the rest of the world may be getting dumber because they have easy access to nonsense. I don't believe the Internet has changed the way I or anyone else thinks. It has changed the arbiters of truth, however. Now everyone is an expert.

Pioneering Insights

Neil Gershenfeld

Physicist; director, MIT's Center for Bits and Atoms; author, Fab: The Coming Revolution on Your Desktop—From Personal Computers to Personal Fabrication

The Internet is many things: good and bad (and worse) business models, techno-libertarian governance and state censors, information and misinformation, empowerment and addiction. But at heart it is the machine with the most parts ever created. What I've learned from the Internet comes not from Web 2.0 or Anything Else 1.0; it's the original insights from the pioneers that made the Internet's spectacular growth possible.

One is interoperability. While this sounds like technological motherhood and apple pie, it means that the Internet protocols are not the best choice for any particular purpose. They are, however, just good enough for most of them, and the result of sacrificing optimality has been a world of unplanned synergies.

A second is scalability. The Internet protocols don't contain performance numbers that impose assumptions about how they will be used, which has allowed their performance to be scaled over six orders of magnitude, far beyond anything initially anticipated. The only real exception to this was the address size, which is the one thing that has needed to be fixed.

Third is the end-to-end principle: The functions of the Internet are defined by what's connected to it, not by how it is constructed. New applications can be created without requiring anyone's approval and can be implemented where information is created and consumed rather than centrally controlled.

A fourth is open standards. The Internet's standards were a way to create playing fields, not score goals; from VHS versus Betamax to

HD-DVD versus Blu-ray, the only thing that has changed in standards wars is who's sitting on which side of the table.

These simple-sounding ideas matter more than ever, because the Internet is now needed more than ever—but in places it has never been. Three-quarters of electricity is used by building infrastructure, which wastes about a third of that, yet many of the attempts to make it intelligent hark back to the world of central office switches and dumb telephones. Some of the poorest people on the planet are "served" by some of the greediest telcos, whereas it is now possible to build communications infrastructure from the bottom up rather than the top down. In these and many more areas, four decades of Internet development are colliding with practices brought to us by (presumably) well-meaning but ill-informed engineers who don't study history as part of an engineering education and thereby doom everyone else to repeat it. I'd argue that we already know the most important lessons of the Internet. What matters now is not finding them but making sure we don't need to keep refinding them.

Thinking in the Amazon

Daniel L. Everett

Dean of Arts and Sciences, Bentley University, Waltham, Massachusetts; author, Don't Sleep, There Are Snakes: Life and Language in the Amazonian Jungle

During the late 1970s and early 1980s, I spent months at a time in the Brazilian Amazon, in complete isolation with the Pirahã people. My only connection with the wider world was a clunky Philips shortwave radio I bought in São Paulo. In the darkness of many Amazonian nights, I turned the volume low and listened, when all the Pirahãs and my family were asleep, to music shows such as *Rock Salad*, to Joan Baez and Bob Dylan, and to news of such events as the Soviet invasion of Afghanistan and the election of Ronald Reagan. As much as I enjoyed my radio, though, I wanted to do more than just listen passively. I wanted to talk. I would lie awake after discovering some difficult grammatical or cultural fact about the Pirahã and feel lost. I could barely wait to ask people questions about the data I was collecting and my ideas about it. I couldn't, though—too isolated. So I put thoughts of collaboration and consultation out of my head. Now, this wasn't a completely horrible outcome; isolation taught me to think independently. But there were times when I would have liked to have had a helping hand.

All that changed in 1999. I bought a satellite phone with Internet capability. I could e-mail from the Amazon! Now I could read an article or a book in the Pirahã village and immediately contact the author. I learned that if you begin your e-mail with, "Hi, I am writing to you from the banks of the Maici River in the Amazon jungle," you almost always get a response. I would send half-baked ideas around the world to colleagues and people I didn't even know and get responses back quickly—sometimes while I was floating down the Maici in my boat, drinking a beer and relaxing from the demands of being the main entertainment for a village of practical-joking Pirahãs. After reading these responses, I would discard some of my ideas, develop others, and, most important, contemplate brand-new ones. I could not have telephoned my interlocutors; most were too busy to take random phone calls from conversation-hungry Amazonianists. And I didn't know most of them all that well. Sending a regular letter was not possible from the Pirahã village. My thinking about language and culture were altered profoundly by access to fresh intellectual energy.

In the city, where I now do most of my work, the Internet has become an extension of my memory—it combats the occasional senior moment, helping me to find names, facts, and places nearly instantly. It gives me a second, bigger brain. The Internet has allowed me to learn from people I've never met. It has placed me in a university that has profoundly affected my career, my research, and my worldview.

I rarely connect to the Internet from the Amazon these days. I'm not there as long or as frequently as in the past. I've learned that the Internet is just a tool. It doesn't fit every job. I avoid using it for tasks requiring a more personal connection, such as administering my university department or talking to my children. But if it's just a tool, it's a wondrous tool. It changed my thinking (and my approach to thinking), just as the first chain saw must have affected loggers. The Internet gave me access to as much information (for good or ill) as any researcher in the world, even from the rain forest.

The Virtualization of the Universe

David Gelernter

Computer scientist, Yale University; chief scientist, Mirror Worlds Technologies; author, Judaism: A Way of Being

The Internet is virtualizing the universe, which changes the way I act and think. Virtualization (a basic historical transition, like industrialization) means that I spend more and more of my time acting within—and thinking about the mirror reflection of—some external system or institution in the (smooth, pondlike) surface of the Internet. But the continuum of the cybersphere will emerge from today's bumpy cob-Web when virtualization reaches the point at which the Internet develops its own emergent properties and systems—when we stop looking at the pixels (the many separate sites and services that make up the Web) and look at the picture. (It's the picture, not the pixels! Eventually top-down thinking will replace bottom-up engineering in the software world—which will entail a roughly 99.9 percent turnover in the current population of technologists.)

Conversation spaces, for example, will be simple emergent systems in the cybersphere, where I talk and listen (or read and write) in a space containing people with whom I like to converse, with no preliminary set-up (as long as there's a computer nearby), as if I were in a room with friends. If I want someone's attention, I say his name or look at him; if I speak a little louder, I'm seeking a general discussion. If I say "Let's talk about Jasper Johns," the appropriate group of people materializes. If one of them is busy, I can speak now and he can speak back to me later, and I can respond later still. Some people claim to be good at multitasking; we'll see how many slow-motion conversations they can keep going simultaneously.

Today there are many universities and courses online; eventually, as virtualization progresses, we'll see many, or most, absorbed into a world university, where you can walk the halls, read the bulletin boards, and peek into classrooms within a unified space without caring which conventional university or Website contributed what. We'll see new types of institutions and objects emerge, too. Virtual objects and institutions will absorb their own histories (like cloth absorbing the fragrance of flowers), so I can visit virtual Manhattan now or roll it backward in time; a large subset of all the knowledge that exists about, say, Wells Cathedral is absorbed into the virtual or emergent Wells Cathedral. At virtual Wells, I can dive deeper for detail about any aspect of the place or roll the building (and its associated ideas and institutions) backward in time until they vanish into the mists of history; or, for that matter, I can tentatively push virtual Wells forward in time (which is not so easy—like pushing something uphill) and see what can be calculated, forecast, or guessed about the cathedral's future a day, a week, or a thousand years from now.

Virtualization has the important intellectual side effect of leading us toward a better understanding of the relation between emergent properties and virtual machines or systems. Thus "I" am an emergent property of my body and mind; "I" (my subjective experience of the world and my self) am a virtual machine, of sorts; but "I" (or "consciousness") am just as real (despite being virtual) as the pull-down menu built of software—or the picture that emerges from the pixels. Like industrialization, virtualization is an intellectual as well as a technological and economic transition; like industrialization, it's a change in the texture of time.

Information-Provoked Attention Deficit Disorder

Rodney Brooks

Panasonic Professor of Robotics, MIT Computer Science and Artificial Intelligence Lab; author, Flesh and Machines: How Robots Will Change Us

When a companion heads to the bathroom during dinner, I surreptitiously pull out my iPhone to check my e-mail and for incoming SMS. When I am writing computer code, I have my e-mail inbox visible at the corner, so that I can see if new messages arrive—even though I know that most that do arrive will be junk that has escaped my spam filters. When I am writing a paper or letter or anything else serious, I flip back and forth, scanning my favorite news sites for new gems; on weekdays I check on stock prices—they might be different from what they were five minutes ago.

I recently realized why I enjoy doing a mindless but timed Sudoku puzzle so much—the clock stops me from breaking off to go graze on the endless variety of intellectual stimulations the Web can bring to me. Tragically, Sudoku is my one refuge from information-provoked attention deficit disorder.

The Internet is stealing our attention. It competes for it with everything else we do. A lot of what it offers is high-quality competition. But unfortunately, a lot of what it offers is merely good at capturing our attention and provides us with little of long-term import—sugar-filled carbonated sodas for our mind.

We, or at least I, need tools that will provide us with the diet Internet, the version that gives us the intellectual caffeine that lets us achieve what we aspire to but doesn't turn us into hyperactive intellectual junkies.

Recently, as reported in *Nature*, an open group of people interested in mathematics (including some of the best currently active mathematicians in the world) used wikis and blogs to come up with a new and elegant proof of the Hales-Jewett theorem in thirty-seven days. The Internet provided a new forum for geographically disparate people to collaborate and contribute new insights, each small and incremental, enabling a result that at best might have taken the brightest of them many months or years to achieve individually.

We can now find just about any scientific paper we want online; I've found some old ones of mine that I had no idea were digitized. I was a smart young thing once, I must say. Soon, just about everything ever written or recorded will be available in some form on the Internet, immediately.

The two promises—ease of collaboration and instant access to any and all information—do indeed change the way we work. Just as Arabic numerals empowered our computation abilities, and just as mass-produced books empowered many more people to have a reference library, and just as the tape recorder and camera empowered us to record data better for careful analysis, and just as calculators and computers empowered us to simulate physical systems without a direct physical analog, the Internet has empowered us to do new and grander things more quickly than previously possible.

But there are kinks yet to be worked out, besides the theft of our attention. There is stability of pointers (so that on our desktop machines our files may move around on the disk but the pointers to them will automagically update to the new location), there is stability of format (so that old movies or documents will still be readable), there is the issue of being able to aggregate digital media into manipulable containers (I used to use cardboard portfolio file cases to organize multiple media for each of my current projects), and then there is that pesky problem of business models, so that people will have a way of getting paid for things they do that we all use.

We're still in the middle of it. We operate in new ways, but those ways have not yet stabilized. Ultimately they will, at least for some of us. I'm hoping I will find my way into that group.

Present Versus Future Self

Brian Knutson

Associate professor of psychology and neuroscience, Stanford University

Like it or not, I have to admit that the Internet has changed both what and how I think.

Consider the obvious yet still remarkable fact that I spend at least 50 percent of my waking hours on the Internet, compared to 0 percent of my time twenty-five years ago. In terms of what I think, almost all my information (e.g., news, background checks, product pricing and reviews, reference material, general "reality" testing, etc.) now comes from the Web. Given the ubiquity and availability of Web content, how could one resist its influence? Although this content probably gets watered down as a function of distance from the source, consensual validation might offset the degradation. Plus, the Internet makes it easier to poll the opinions of trusted experts. So, overall, the convenience and breadth of information on the Internet probably help more than hurt me.

In terms of how I think, I fear that the Internet is less helpful. Although I can find information faster, that information is frequently tangential. More often than I'd like to admit, I sit down to do something and then get up bleary-eyed hours later, only to realize that my task remains undone (or I can't even remember the starting point). The sensation is not unlike walking into a room, stopping, and asking yourself, "Now, what was I here for?"—except that you've just wandered through a mansion and can't even remember what the entrance looked like.

This frightening, face-sucking potential of the Web reminds me of conflicts between present and future selves first noted by ancient Greeks and Buddhists and poignantly elaborated by philosopher

Derek Parfit. Counterintuitively, Parfit considers present and future selves as different people. By implication, with respect to the present self the future self deserves no more special treatment than anyone else does. Thus, if the present self doesn't feel a connection with the future self, then why forgo present gratification for someone else's future welfare?

Even assuming that the present self does feel connected to the future self, the only way to sacrifice something good now (e.g., reading celebrity gossip) for something better later (e.g., finishing that term paper) is to slow down enough to appreciate that connection, consider the conflict between present and future rewards, and decide in favor of the best overall course of action. The very speed of the Internet and convenience of Web content accelerate information search to a rate that crowds out reflection, which may bias me toward gratifying the salient but fleeting desires of my present self. Small biases, repeated over time, can have large consequences. For instance, those who report feeling less connected to their future self also have less money in their bank accounts.

I suspect I am not the sole victim of Internet-induced present-self bias. Indeed, Web-based future-self prostheses have begun to emerge, including software that tracks time off task and intervenes, ranging from posting reminders through blocking access and even shutting programs down. Watching my own and others' struggles between present self and future self, I worry that the Internet may impose a "survival of the focused," in which individuals gifted with some natural ability to stay on target, or who are hopped up on enough stimulants, forge ahead while the rest of us flail helplessly in a Web-based attentional vortex. All of this makes me wonder whether I can trust my selves on the Internet. Or do I need to take more draconian measures—for instance, leaving my computer at home, chaining myself to a coffeehouse table, and drafting in longhand?

I Am Realizing How Nice People Can Be

Paul Bloom

Psychologist, Yale University; author, How Pleasure Works: The New Science of Why We Like What We Like

When I was a boy, I loved the science fiction idea of a machine that could answer any factual question. It might be a friendly robot, or a metal box you keep in your house, or one of the components of a starship. You would just ask, "Computer, how far away is Mars?" or "Computer, list the American presidents in order of height," and a toneless voice would immediately respond.

I own several such machines right now, including an iPhone that fits in my pocket; all of them access information on the Internet. (Disappointingly, I can't actually talk to any of them—the science fiction writers were optimistic in this regard.) But the big surprise is that much of this information is not compiled by corporations, governments, or universities. It comes from volunteers. Wikipedia is the best-known example, with millions of articles created by millions of volunteer editors, but there are also popular sites such as Amazon.com and TripAdvisor.com that contain countless unpaid and anonymous reviews.

People have wondered whether this information is accurate (answer: mostly yes), but I'm more interested in its very existence. I am not surprised by the scammers, the self-promoters, and the haters. But why do people devote their time and energy to anonymously donating accurate and useful information? We don't put $20 bills in strangers' mailboxes; why are we giving them our time and expertise? Comments on blogs pose a similar puzzle, something nicely summarized in the classic *xkcd* cartoon in which someone is typing frantically on the computer; when asked to come to bed, the person says, “I can’t. This is important. . . . Someone is wrong on the Internet.”

Apparently the Internet evokes the same social impulses that arise in face-to-face interactions. If someone is lost and asks you for directions, you are unlikely to refuse or to lie. It is natural, in most real-world social contexts, to offer an opinion about a book or movie you like or to speak up when the topic is something you know a lot about. The proffering of information on the Internet is the extension of this everyday altruism. It illustrates the extent of human generosity in our everyday lives and also shows how technology can enhance and expand this positive human trait, with real beneficial results. People have long said that the Web makes us smarter; it might make us nicer as well.

My Perception of Time

Marina Abramović

Artist

Since I started using the Internet and all the options it offers in matters of communication, my perception of global time has changed radically.

I'm now much more aware of time differences, and, in a restless way, my nights are haunted by the presence of the other working days all around the world.

I've become obsessed with being constantly up to date on my correspondence, and I've lost that no-man's-land that was the time it took for a letter to arrive at its destination, be answered, and travel all the way back to me.

My days become nights, and my nights become brighter and more "available."

Ever since I understood this trap, I have been trying to fight back, to take back control of my time, but it's hard to do, especially when my perception of time itself has altered.

The Rotating Problem, or How I Learned to Accelerate My Mental Clock

Stanislas Dehaene

Neuroscientist, Collège de France, Paris; author, Reading in the Brain

Like the Gutenberg press in its time, the Internet is revolutionizing our access to knowledge and the world we live in. Few people, however, pay attention to a fundamental aspect of this change: the shift in our notion of time. Human life used to be organized in inflexible day-and-night cycles—a quiet routine that has become radically disrupted, for good or ill.

Some years ago, I was working out of Paris with colleagues in Harvard on the mathematical mind of Amazon Indians. The project was so exciting, and we were so motivated by the paper we were writing, that we worked on it every day, if not day and night (we had families and friends . . .).

At the end of each day, I would send my colleagues a new draft of our article, full of detailed questions and issues that needed to be addressed. In a world without the Internet, I would have had to wait several weeks for a reply. Not so in today's world. Every morning, after a good night's sleep, I woke up to find that most of my questions had been answered during the night, as if by magic. The experience reminded me of the mysterious instances of unconscious problem solving during sleep, as famously reported by Kekulé, Poincaré, Hadamard, and other mathematicians and scientists. The difference, of course, was that my problems were solved thanks to conscious effort and the pooling of several minds around the planet.

For my Harvard colleagues, too, the experience seemed somewhat miraculous. They, too, had many questions, and I dutifully computed the statistics they requested, drew the new data plots they asked for, and wrote the paragraphs they needed—all this while Harvard was still plunged in night. Thanks to this collective effort, our work was completed much faster than any one of us could have managed alone. We had almost doubled the speed of our mental clocks!

The idea is now commonplace. A great many companies outsource translation or maintenance to Indian, Australian, or Taiwanese employees on the other side of the world so that the work can be completed overnight. However, the scope of this phenomenon does not appear to have fully dawned on us yet.

For example, imagine an international corporation—say, a movie studio, such as Pixar—intentionally placing three of its computing centers at the vertices of a giant equilateral triangle spanning the earth, so that the employees at a given location can work on a project for eight daylight hours and then pass it on to another team in a different time zone.

For a more grandiose picture, one that could have arisen from the mind of Jorge Luis Borges, imagine a complex Problem that moves around the planet via the Internet, at a fixed speed precisely countering the Earth's rotation, in such a way that the Problem constantly faces the sun. As dawn comes for a fraction of humanity, the Problem is present on their computer screens—but some of it has been chipped away by armies of fellow workers who, by this time, are sound asleep. Day and night, without interruption, the Earth's rotation cranks away at the Problem until it is solved.

Such giant utopian or Borgesian projects do in fact already exist: Wikipedia, Linux, SourceForge, OLPC (One Laptop per Child). They are beyond the scope, or even the imagination, of any single human being. Nowadays open-source development moves around in the infosphere and is being improved constantly on whatever side of the planet happens to be in sunshine (and often on the other side as well).

There is grandeur in this new way of computer life, where the normal sleep-wake cycle is replaced by the constant churning of silicon and mind. But there is much inherent danger in it as well. Take a look at Amazon's aptly named Mechanical Turk, and you'll find an alternative Website where largely profitable enterprises in developed countries offer short-term, badly paid computer jobs to the third world's poor. For a few pennies, they propose a number of thankless assignments ironically called "human intelligence tasks" that require completing forms, categorizing images, or typing handwritten notes—anything computers still cannot do. They provide no benefits, no contract, no guarantees, and ask no questions: the dark side of the intellectual globalization now made possible by the Internet.

As our mental clocks keep accelerating and we become increasingly impatient about our unfinished work, the Internet provides our society with a choice that deserves reflection: Do we aim for ever faster intellectual collaboration or for ever faster exploitation that will allow us to get a good night's sleep while others do the dirty work? With the Internet, a new sense of time is dawning, but our basic political options remain essentially unchanged.

I Must Confess to Being Perplexed

Mihaly Csikszentmihalyi

Psychologist; Director, Quality of Life Research Center, Claremont Graduate University; author, Flow: The Psychology of Optimal Experience

Answering this question should be a slam dunk, right? After all, thinking about thinking is my racket. Yet I must confess to being perplexed. I am not even sure we have good evidence that the way humans think has been changed by the advent of the printing press. . . . Of course, the speed of accessing information and the extent of information at one's fingertips have been increased enormously, but has that actually affected the way thinking unfolds?

Relying on my personal experience, I would suggest the following hypotheses:

1. I'm less likely to pursue new lines of thought before turning to the Internet to either check existing databases or ask a colleague directly. (Result: less sustained thought?)
2. Information from the Internet is often decontextualized, but, being quickly acquired, it satisfies immediate needs at the expense of deeper understanding. (Result: more superficial thought?)
3. At the same time, connection between ideas, facts, et cetera, can be more easily established on the Web—if one takes the time to do so. (Result: more intrapersonally integrated thought?)
4. The development of cooperative sites, ranging from Wikipedia to open-source software (and including *Edge*?), makes the thought process more public, more interactive, more transpersonal, resulting in something similar to what Teilhard de Chardin anticipated over half a century ago as the

noosphere, a global consciousness he saw as the next step in human evolution.

Like all technologies, this one has both positive and negative consequences. I'm not sure I'd bet on the first two (negative) hypotheses being closer to the truth—or on the next two, which are more positive. And of course, both sets could be true at the same time.

Taking on the Habits of the Scientist, the Investigative Reporter, and the Media Critic

Yochai Benkler

Berkman Professor of Entrepreneurial Legal Studies, Harvard Law School; codirector, Berkman Center for Internet and Society; author, The Wealth of Networks: How Social Production Transforms Markets and Freedom

Answering this question requires us to know what you mean by "the Internet" and what you mean by "the way you think." If by "the way you think," you mean "the way your brain functions when you are doing certain kinds of operations," I am provisionally prepared to answer, "Not at all." Provisionally, because it is not entirely clear to me that this is true.

I will answer the question as though it were phrased, instead, as "How has the Internet changed the way you come to form and revise beliefs?"—beliefs about the state of the world (for example, that the globe is warming due to human action) or the state of a social claim (whether or not a blue shirt goes with black pants, or whether it is immoral to enforce patents on medicines in ways that result in prices too high for distribution in Africa, where millions of people die each year from preventable diseases while manufacturers of generics stand ready to make those drugs affordable).

This leads us to the first question: What do you mean by "the Internet"? By "Internet" I, at least, mean a sociocultural condition in which we are more readily and seamlessly connected to more people, with varying degrees of closeness and remoteness; to more social and organizational structures, both those we belong to and those we don't; and to more cultural artifacts and knowledge-embedded objects.

An e-mail with an inchoate thought, half a fragment to a friend, is the kind of thing I can do today with more people than those with whom I can readily grab a cup of coffee; I can also do it with people whose friendship I value but who are geographically remote. Social distance has moderated as well. Sending an e-mail to a stranger who stands in an organizational, institutional, or socially proximate role is slightly easier and considered less intrusive than making a phone call used to be.

Most radical is the recognition that someone, somewhere, entirely remote in geographic, social, or organizational terms, has thought about something similar or pertinent. Existing as we do in a context that captures the transcript of so many of our conversations—from Wikipedia to blogs—makes the conversations of others about questions we are thinking about vastly more readable to us than was true in the past.

If by "the way I think" we evoke Descartes' *cogito*, the self-referential "I think," then all we would think of with regard to the Internet is information search and memory enhancement. But if we understand thought as a much more dialogic and dialectic process—if "I think" entails "I am in conversation"—then the Internet probably does change how I think, quite a bit. No, it doesn't mean that "everyone is connected to everyone else" and we all exist in a constant stream of babble. But it does mean that we can talk to one another in serially expanding circles of social, geographic, and organizational remoteness, and that we can listen to others' conversations, and learn.

Thinking with these new capabilities requires both a new kind of open-mindedness and a new kind of skepticism. Open-mindedness, because it is increasingly turning out that knowledge and insight reside in many more places than we historically recognized. A sixteen-year-old Norwegian kid might solve the question of how to crack the DVD scrambling system. A ski lift operator and shoe salesman from Minnesota who happens to be a political junkie and hangs out on *Daily Kos* may have more insights into the dynamics of the Minnesota Senate election recount than the experts at CNN or the *New York Times.*

But there is also plenty of nonsense. We all know this. And so alongside the open-mindedness, we have come to develop a healthy dose of skepticism—both about those who are institutionally anointed experts and about those who are institutional outsiders. Belief formation and revision is an open and skeptical conversation: searching for interlocutors, forming provisional beliefs, giving them weight, continually updating. We cannot seek authority, only partial degrees of provisional confidence. We have to take on the habits of the scientist, the investigative reporter, and the media critic as an integral part of the normal flow of life, learning, and understanding.

Maybe that's how I've always been. Maybe it has nothing to do with the Internet.

Thinking as Therapy in a World of Too Much

Ernst Pöppel

Psychologist and neuroscientist; director emeritus, Institute for Medical Psychology, Ludwig-Maximilians-Universität München; author, Mindworks: Time and Conscious Experience

It is painful to admit, but I had never thought about thinking before the Internet. With a prescientific attitude, I had (and most of the time still have) the impression that "I" do not think at all, but that "it thinks," sometimes resulting in what appears to be a solution or an insight but usually ending in nowhere. Apparently I am at the mercy of uncontrolled and uncontrollable processes, presumably in my brain, but before the Internet I never cared about these processes themselves.

This is how I experience this "proto-thinking": It is like swimming in an ocean with no visible horizon, where sometimes an island surfaces unexpectedly, indicating a direction, but before I reach this island it has disappeared again. This feeling of being at a loss has become much stronger with the Internet. There is no direction, there are no islands—and this can no longer be accepted. What can I do, swimming in this ocean of information, in a world of too much? Maybe it is useful to think about thinking—as others such as John Brockman have done successfully, which then enables them to ask a question about thinking and the Internet. Maybe it is helpful to think about thinking as a therapy to combat loss of cognitive control—to fight against the "too much" that results in "too little." The goal must be to create a personal framework for orientation in the world of too much, by asking questions such as "What is thinking?" or "Why is there thinking?"

These are my personal answers, presumably shared by many others. Why is there thinking? From a biological point of view (and can there be any other?), thinking is a service function of our brain to create a homeostatic state, an internal equilibrium. Of course, thinking is not the only service function; this is also true of perception, emotional evaluation, and working memory. But those functions are characterized by a rather short time horizon. To expand this horizon, thinking arrived in our evolution; thus, virtual behavior has become possible. Goal-oriented thinking allows the anticipation of a successful action and creates freedom in behavioral control such that the organism no longer has to react instantaneously. The option space of potential successful actions to reach a homeostatic state is considerably enlarged.

Next question: What is thinking? For successful *Probehandeln* (as Sigmund Freud referred to thinking), the letter *C* may serve to remind us of the different operations.

Thinking is necessarily defined within a *context*, or frame. Without context, I navigate the Internet without any orientation, hoping to find a jewel by harvesting serendipity (which, indeed, sometimes happens).

Thinking requires material for thinking operations. That is, without a mental *category*, there would be no thinking. Thinking must be *about* something, clearly and distinctly defined, as René Descartes asked for in his first rule of thinking in the *Discours de la méthode*. But one category is not sufficient.

Thinking requires several categories in order to allow *comparison*, which is, according to Rudolf Carnap, the most basic mental operation. Comparing is possible with respect to quantity or to quality. That is, different categories can be more or less, or "this versus that," and the result of a comparison allows *choice*, which is the basis of decision and then of action.

The process of categorizing, comparing, and choosing must follow a correct temporal order or sequence, which only then allows the extraction of *causality*, based on the proper *continuity* of mental operations. But how do I know whether thinking has brought me to the right answer?

The *constellation* of the different operations, and the answer gained by thinking about a question, has to fit into the landscape of previous thinking and what is considered to be true. This may be signaled by what Archimedes experienced as "Eureka!" This experience is more than an analytical appreciation; it results in a feeling of satisfaction that indeed the anticipated goal in virtual space has been reached.

Certainly I never would have thought about the seven C's as elements of thinking if I hadn't been lost in the world of too much. Thus, thinking has become a necessary therapy.

internet is wind

Stefano Boeri

Architect, Politecnico of Milan; visiting professor, Harvard University Graduate School of Design; editor-in-chief, Abitare *magazine*

internet is wind.

a constant—and dominant—wind that unsettles and swathes us.

in recent years we have become familiar with walking by displacing our weight, our equilibrium, in an opposite direction to this wind.

only in this manner are we able to walk straight, without succumbing, without completely folding to its logic of simultaneous and globalized reciprocity. but it is enough to unplug the connection, turn the corner, find shelter, place oneself leeward, and internet disappears.

leaving us unbalanced, for a moment, folded in the direction of the wind, because of the inertia of the effort of resistance we have made until that moment.

and yet, at that moment, the effort seems a formidable resource.

suddenly we are in front of what is not said; of that which we cannot and will not ever communicate of our own interior, of our personal idiosyncrasies, of our distorted individuality.

thought in the era of internet has this uniqueness:

there, the space-time we are able to protect from this wind becomes precious occasions to understand what we cannot say, what we are not willing to deposit in the forum of planetary simultaneity.

so as to understand what we really are.

Of Knowledge, Content, Place, and Space

Galia Solomonoff

Architect, Solomonoff Architecture Studio

The Internet is fundamentally altering the relationship between knowledge, content, place, and space. Consider the world as divided into two similarly populous halves: people born before 1980 and people born after 1980. Of course, there are other important differences, such as gender, race, class, ethnicity, and geography, yet I see 1980 as significant in the shift and alteration in the relationship of knowledge, place, and space—because of the use of the Internet. Three examples/scenes:

Example/Scene 1

I am responding to the *Edge* question from Funes, a locality of 15,000 inhabitants in the middle of the Argentine Pampas. I am in a *locutorio* with eight fully equipped computers that charges the equivalent of twenty cents for fifteen minutes of Internet use. Five other users are here: a woman in her twenties talking via Skype to her sister and niece in Spain, a thirtysomething man in shirt and tie scanning a résumé, two teens playing a video game with what I guess is a multiplaced or nonplaced community, and a man posting photos of a baby and a trip on a Facebook page. And there's me, a forty-two-year-old architect on vacation, with an assignment due in two hours!

I am the oldest here. I am also the only nonlocal. The computer helps me and corrects my spelling without being prompted.

Example/Scene 2

Years ago, when I was an architecture student and wanted to know about, say, Guarino Guarini's importance as an architect, I would go two flights down at Avery Library, get a few index cards, follow the numbered instructions on them, and find two or four or seven feet of books on a shelf dedicated to the subject. Then I would look at a few cross-referenced words on those cards, such as "Mannerist architecture," go down another aisle in the same room, and identify another few feet of books. I would leaf through all the found books and get a vague yet physical sense of how much there was to know about the subject matter.

Now I Google "Guarino Guarini" and in 0.29 seconds I get 153,000 entries. The first page reveals basic details of his life: He was born on January 7, 1624, and lived until March 6, 1683. I also get six images of cupolas, a Wikipedia entry, and an *Encyclopaedia Britannica* entry. My Google search is very detailed, yet not at all physical. I can't tell how much I like this person's personality or work. I can't decide whether or not I want to flip through more entries.

Example/Scene 3

I am in a car traveling from New York to Philadelphia. I have GPS but no maps. The GPS tells me where to go and takes into account traffic and tolls. I trust the GPS, yet in my memory I wish to reconstruct a New York–Philadelphia trip I took years ago. On that other trip, I had a map, I entered the city from a bridge, the foreground was industrial and decrepit, the background was vertical and contemporary . . . At least, that is what I remember. Was it so? I zoom out the GPS to see if the GPS map reveals an alternative entry route, a different way the city geography could be approached. Nothing in the GPS map looks like the space I remember. What happened? Is my memory of the place faulty, or is the focus of the GPS too narrow?

What I want to convey with these examples/scenes is how, over time and with the advent of the Internet, our sense of orientation, space, and place has changed, along with our sense of the details necessary to make decisions. If decisions take into account the many ways in which information comes to us, then the Internet at this point privileges what we can see and read over many other aspects of knowledge and sensation, such as how much something weighs, how it feels, how stable it is. Are we, the ones who knew places before the Internet, more able to navigate them now, or less? Do we make better or worse decisions based on the content we take in? Do we have longer, better sojourns in faraway places or constant placelessness? How have image, space, place, and content been altered to give us a sense of here and now?

The Power of Conversation

Gloria Origgi

Philosopher, Institut Jean Nicod, Paris

I spend more than half my working hours doing e-mail. I have 4,407 messages in my Gmail inbox today—stuff that I haven't read yet, that I have to reply to, or that I keep in the inbox just so I can take advantage of the search function and retrieve it when needed.

Each time I find myself late in the afternoon still writing messages to friends, colleagues, perfect strangers, students, et cetera, I feel guilty at having wasted my day, shirked my intellectual responsibility. These psychological reactions can be harsh to the point of forcing me to inflict various forms of punishment on myself, such as imprisonment in a dusty Parisian library without Internet connection or switching off the modem at home. That is because I believe that my work is *not* writing e-mails; rather, it is writing papers and learned essays on philosophy and related issues.

But what is philosophy? What is academic work in general, at least in the humanities? One of my mentors once told me that being an academic just means being part of a conversation—that's it. Plato used the dialog as a form of expression, to render in a more vivid way the dialectic process of thinking, to construct knowledge from open verbal confrontation. One of the books that influenced me most during my undergraduate philosophical studies in Italy was Galileo's *Dialogue Concerning the Two Chief World Systems*. I have read on the *Edge* site that *Edge* is a conversation. So—what is so bad about the e-mail conversations invading my life? What is the big difference between contemplating the first blank page of a new paper and the excited exchange through Gmail or Skype with a colleague in another part of the world?

My intellectual life started to improve when I realized there isn't that much difference—that academic papers, comments on papers, reviews, replies, et cetera, are conversations in slow motion. I write a paper for an academic journal; the paper is evaluated by other philosophers, who suggest improvements; it is then disseminated to the academic community to prompt new conversations on a topic or launch new topics for discussion. Those are the rules of the game. And if I make an introspective effort to visualize my way of thinking, I realize I am never alone in my mind: A number of more or less invited guests are sitting around somewhere in my brain, challenging me when I overconfidently claim this or that.

Arguing is a basic ingredient of thinking: Our ways of structuring thought would have been very different without the powerful tool of verbal exchange. So let's admit that the Internet allows us to think and write in a way much more natural than the one imposed by the traditional culture of the written word. The dialogical dimension of our thinking is now enhanced by continual fluid exchanges with others.

The way to stop feeling guilty about wasting our time is to commit ourselves to interesting and well-articulated conversations—just as we accept invitations to dinners where we hope the talk will be stimulating and we won't fall asleep after the second glass of wine. I run a Website that keeps track of high-level learned conversations between academics. I find that each medium produces its wastes; for example, most books are just noise that disappears a few months after publication. I don't think we should concentrate on the wastes; rather, we should make responsible use of our conversational skills and free ourselves from unreal commitments to accidental formats, such as books or academic papers—formats that owe their authority to the central role they played in our education.

If what we leave to the next generation are threads of useful and learned conversations, then so be it. I see this as an improvement in our way of externalizing our thinking—a much more natural way of being intelligent in a social world.

A Real-Time Perpetual Time Capsule

Nick Bilton

Adjunct professor, Interactive Telecommunications Program, New York University; lead technology writer, Bits *blog, the* New York Times; *author,* I'm from the Future and Here's How It Works.

The Internet has become a real-time perpetual time capsule. A bottomless invisible urn. A storage locker for every moment of our lives, and a place to allow anyone to dip in and retrieve those memories.

The Internet has killed the private diary hiding under my sister's mattress and replaced it with a blog or social network.

Through the social sharing Web, we have become an opt-everything society: sharing our feelings in status updates, uploading digital pictures of everything—good or otherwise. We discuss what we're reading or watching and offer brutally honest critiques. We tweet the birth of a child or announce an engagement. And we are completely unaware of the viewers we talk with. I suspect we don't even care. (I know I don't.) We are all just part of an infinite conversation.

And no one stands above anyone else. The Internet gives each of us a bullhorn and allows us to use it freely, whenever we see fit, to say whatever we want. In the past, bullhorns were expensive, as were printing presses, television studios, or radio stations. To reach large audiences required deep pockets. But now we are all capable of distributing our voices, opinions, and thoughts evenly. When everyone has a bullhorn, no one individual can shout louder than the others; instead, it just becomes a really loud conversation.

Most important, the Web allows for an equilibrium of chatter. People use the same services to share and consume their vastly divergent views and interests.

The Web is capable of spreading information more quickly than any virus known to man, and it's impossible to stop. Without these confabulations, the Web would be an empty wasteland of one-sided conversation, just as newspapers, television programs, and radio stations used to be.

The Internet has changed the way we think, through numerous channels. But it has changed the way I think through one very simple action: Every important moment of my life is documented, cataloged, and sent online to be shared and eulogized with whoever wants to engage in the conversation.

Getting from Jack Kerouac to the Pentatonic Scale

Jesse Dylan

Filmmaker; founder, FreeForm production company; founder, the medical Website Lybba.org

The promise of the Web, when it was first kicked around at CERN and DARPA, was to create a decentralized exchange of information. The grand power of that idea is that insight can come from anywhere. People with differing ideas and backgrounds can test their theories against the world, and may the best idea win. The fact that the information can be looked at by so many different kinds of people from anywhere on Earth is the Internet's true power—and the source of my fascination with it. Right now, a little kid can browse the raw data coming from the Large Hadron Collider; he can search the stars for signals of alien life with the SETI project. Anyone can discover the next world-changing breakthrough. That's the point of the Internet.

Also, the contribution of search engines in simplifying the research process can't be underestimated. This enables us to conduct research instantly, on our own terms. That's a tremendous leap from what I had to do ten years ago to find anything out—from knowing who my interview subjects are to where I can get the best BLT in Hollywood—and the Web is still in its infancy. The great hubs of information we've constructed, and the tools to traverse them, such as Google, Wikipedia, and Facebook, are only going to get deeper and more resonant as we learn how to communicate over them more effectively. Just think about what we can do when these tools are applied to the worlds of medicine, science, and art. I can't wait to see what a world full of instant knowledge and open inquiry will bring.

Today, the Internet permeates pretty much all of my thoughts and actions. I access it with my phone, my computer, at home, at work. It gives me untold quantities of new knowledge and inspiration. I interact with people all over the world from different fields and walks of life, and I see myself and others becoming interconnected hubs of information that the full range of human experience passes through. I feel that I am never truly alone, with the ends of the Earth a few clicks away.

I once discussed with the late mathematician George Whitehead the way to approach innovation. Almost as an aside, he said that the only way to make advances was to have five different strategies, in the hope that one of them would work out. Well, the Internet is a place where I can pick from the sum of all the strategies people have already tried out, and if I think of something new, I can put it up there to share with the world.

I was at the Mayo Clinic doing a film project on a rare condition called neuromyelitis optica (NMO, aka Devic's disease or Devic's syndrome). I heard about how the diagnostic test for this condition was discovered. A specialist on multiple sclerosis was speaking at a symposium, and a cancer researcher heard his results. This accident led to the creation of the test. To me, though, it's not an accident at all. It happened because someone—maybe the Mayo brothers themselves—put in place a system, making the symposium an event that disparate researchers and physicians would attend. The insight came because the platform made it possible for these people and ideas to come together, and that made possible a better level of understanding, and so on and so forth.

When I was a child, I learned from looking at the world and reading books. The knowledge I craved was hidden away. Much was secret and unavailable. You had to dig deep to find what you were looking for, and often what you wanted was locked up and out of reach. To get from Jack Kerouac to Hank Williams to the pentatonic scale used to be quite a journey. Now, it can happen in an instant. Some people would say that the old way was a good thing. I disagree.

A Vehicle for Large-Scale Education About the Human Mind

Mahzarin R. Banaji

Richard Clarke Cabot Professor of Social Ethics, Department of Psychology, Harvard University

My first encounter with the information highway came in the form of a love letter in 1982. My boyfriend had studied artificial intelligence at Carnegie Mellon in the mid-1970s and worked at an IBM lab on the East Coast while I was in graduate school in the Midwest. He had pestered me to get an account on something called BITNET. After procrastinating, because I didn't see the point of it, there I was, connected to him without paying AT&T a penny. So that's what the Net is good for, I thought, and recommended it wholeheartedly to every couple struggling to manage a long-distance relationship.

Almost thirty years later, I cannot say that the Internet has changed, even an iota, how I think. But what the Internet has surely done is to change what I think about, what I know, and what I do. It has done so in stupendous ways, and I mention the single most significant one.

In the mid-1990s, I began working on a method for gaining access to the way in which the mind works automatically, unreflectively, less consciously. My students and I studied how thoughts and feelings about social groups (race, gender, class, age, and so on)—feelings we might consider unacceptable—nevertheless came to have a presence in our minds. This situation, we recognized, didn't result from simple obtuseness on the part of human beings; it was the mind's nature. Remarkably, I could test myself, and I learned that my own mind contained thoughts and feelings of which I was unaware, that those thoughts and feelings weren't ones I wanted to possess or was proud of, yet much as I might deny them, they were a part of who I was.

In 1998, my collaborator Tony Greenwald and I decided it was time to develop a version of the test—called the Implicit Association Test, or IAT—for the Web. There were no models for doing this; there were no such experiments by behavioral scientists at the time. But we had talent and grit in the person of Brian Nosek, a graduate student at Yale; a visionary in Phil Long, Yale's main IT overseer; and a scrupulous and effective Internal Review Board that worked through the ethical details of such a presence on the Web.

We went live on September 29, 1998, agreeing that our main purpose for placing the IAT on the Internet was not research as much as it was education. We believed that the method we had developed could provide a moment of self-reflection and learning. That if we did it right, we could engage thousands, even millions, in the task of asking where the stuff in their heads comes from, in what form it sits there, and what they may want to do about it if they don't approve of it.

In the very first days, a large news network placed a link to our site, and there was no looking back. Hundreds of people visited, sampled the IAT, and fired off their responses at us. Interactions with them about technical issues, but even more so about their reactions to the experience, forced us to write new language and modify our presentation. By the end of the first month, we were the stunned recipients of 40,000 completed IATs. We couldn't have learned what we did in that month in half a lifetime had we stayed with the traditional platform for research.

This site, whose primary purpose was educational, changed the research enterprise itself. A research question involving an alternative hypothesis posed on day one could be answered by day two because of the amount of data that flowed in. The very nature of research changed—in the collaborations that mushroomed, in the diversity of the people who participated, in the sheer amount we were able to learn and know at high speed.

The Internet has changed the quality of what we know and increased our confidence in our assessments of what we know. It has changed our notion of what it means to be in constant public dialog about our science. It has changed our relationships with project participants, with whom there can be a real discussion, sometimes many months after the initial interaction. It has also changed our relationship with the media, whose practitioners became research subjects themselves before communicating about the work. Most surprising was the discovery that the vast majority of visitors to the site were willing to entertain the notion that they might not know themselves. Without the Internet, we might have believed that this was the limited privilege of the intellectual elite. Now we know better.

Of course, this science will always require other forms of gathering data besides the Internet. Of course, there are serious limits to what can be done to understand the human mind using the vehicle of the Internet. But it is safe to say that the Internet allowed us to perform the first large-scale study of an aspect of social cognition. Today we have more than 11 million pieces of IAT data from implicit.harvard.edu and its predecessor site. The topics cover what the site is best known for (automatic attitudes toward age, race/ethnicity, sexuality, skin color, religion, automatic stereotypes of foreignness, math/science, career/home) as well as political attitudes in the last three presidential elections and dozens of matters concerning health, mental health, consumer behavior, politics, medical practice, business practice, legal matters, and educational interests. Any person with access to the Net and a desire to spend a few minutes locked in battle with the IAT is a potential participant in the project. Teachers and professors, corporations and nonprofits all over the world use the site for their own educational purposes.

The site yields 20,000 completed IATs per week and involves specialized sites for thirty-three countries in twenty-two languages. There are no advertisements. Somehow, people find it, and stay. They stay, we think, for the simple reason that they want to understand themselves better.

Sandbars and Portages

Tim O'Reilly

Founder and CEO of O'Reilly Media, Inc.

Many years ago, I began my career in technology as a technical writer, landing my first job writing a computer manual on the same day I saw my first computer. The only skill I had to rely on was one I acquired in my years as a reader and my university training in Greek and Latin classics: the ability to follow the bread-crumb trail of words back to their meaning.

Unfamiliar with the technology I was asked to document, I had to recognize landmarks and connect the dots—to say, "These things go together." I would read a specification written by an engineer over and over until I could read it like a map and put the concepts in the right order, even if I didn't yet fully understand them. That understanding would come only when I followed the map to its destination.

Over the years, I honed this skill, and when I launched my publishing business the skill I had developed as an editor was that of seeing patterns: "Something is missing here"; "These two things are really the same thing, seen from different points of view"; "These steps are in the wrong order"; "In order for *X* to make sense, you first have to understand *Y*." Paula Ferguson, one of the editors I hired, once wrote that "all editing is pattern matching." You study a document, and you study what the document is talking about, and you work on the document until the map matches the territory.

In those early years of trying to understand the industry I'd been thrust into, I read voraciously, and it was precisely because I didn't understand everything I read that I became skilled at recognizing patterns. I learned not as you are taught in school, with a curriculum and a syllabus, but as a child explores and composes a worldview bit by bit out of the stuff of everyday life.

When you learn in this way, you draw your own map. When my co-worker Dale Dougherty created GNN—the Global Network Navigator, the first commercial Web portal—in 1993, he named it after *The Navigator*, a nineteenth-century handbook that documented the shifting sandbars of the Mississippi River. My own company has been a mapmaker in the world of technology—spotting trends, documenting them, and telling stories about where the sandbars lie and where the portages that cut miles off the journey are, along with the romance of travel and the glories of the destination. In telling stories to explain what we've learned and encouraging others to follow us into the West, we've become not just mapmakers but meme makers. Open source, Web 2.0, the Maker movement, government as a platform—all are stories we've had a role in telling.

It used to be the case that there was a canon, a body of knowledge shared by all educated men and women. Now we need the skills of a scout: the ability to learn, to follow a trail, to make sense out of faint clues, to recognize the way forward through the thickets. We need a sense of direction that carries us onward despite our twists and turns. We need "soft eyes" that take in everything we see, not just what we are looking for.

The information river rushes by. Usenet, e-mail, the World Wide Web, RSS, Twitter—each generation carrying us faster than the one before.

But patterns remain. You can map a river as well as you can map a mountain or a wood. You just need to remember, the next time you come by, that the sandbars may have moved.

No One Is Immune to the Storms That Shake the World

Raqs Media Collective

Artists, media practitioners, curators, editors, and catalysts of cultural processes

We are a collective of three people who began thinking together almost twenty years ago, before any of us ever touched a computer or logged on to the Internet.

In those dark days of disconnect, in the early years of the final decade of the last century in Delhi, we plugged into each other's nervous systems by passing a book from one hand to another, by writing in each other's notebooks. Connectedness meant conversation. A great deal of conversation. We became each other's databases and servers, leaning on each other's memories, multiplying, amplifying, and anchoring the things we could imagine by sharing our dreams, our speculations, and our curiosities.

At the simplest level, the Internet expanded our already capacious, triangulated nervous system to touch the nerves and synapses of a changing and chaotic world. It transformed our collective ability to forage for the nourishment of our imaginations and our curiosities. The libraries and archives we had only dreamed of were now literally at our fingertips. The Internet brought with it the exhilaration and abundance of a frontierless commons, along with the fractious and debilitating intensity of depersonalized disputes in electronic discussion lists. It demonstrated the possibilities of extraordinary feats of electronic generosity and altruism, with people sharing enormous quantities of information on peer-to-peer networks, and at the same time it provided early exposure to, and warnings about, the relentless narcissism of vanity blogging. It changed the ways in which the world became present to us and the ways in which we became present to the world, forever.

The Internet expands the horizon of every utterance or expressive act to a potentially planetary level. This makes it impossible to imagine a purely local context or public for anything that anyone creates today. It also decenters the idea of the global from any privileged location. No place is any more or less the center of the world than any other anymore. As people who once sensed that they inhabited the intellectual margins of the contemporary world simply because of the nature of geopolitical arrangements, we know that nothing can be quite as debilitating as the constant production of proof of one's significance. The Internet has changed this one fact comprehensively. The significance, worth, and import of one's statements are no longer automatically tied to the physical fact of one's location on a still unequal geopolitical map.

Although this does not mean that, as artists, intellectuals, and creative practitioners, we stop considering or attending to our anchorage in specific coordinates of actual physical locations, it does mean we understand that the concrete fact of our physical place in the world is striated by the location's transmitting and receiving capacities, which turn everything we choose to create into either a weak signal or a strong one. We are aware that these signals go out not just to those we know and who know us but to the rest of the world, through possibly endless relays and loops.

This changes our understanding of the public for our work. We cannot view our public any longer as being arrayed along familiar and predictable lines. The public for our work—for any work that positions itself anywhere vis-à-vis the global digital commons—is now a set of concentric and overlapping circles, arranged along the ripples produced by pebbles thrown into the fluid mass of the Internet. Artists have to think differently about their work in the time of the Internet, because artistic work resonates differently and at different amplitudes.

More often than not, we are talking to strangers on intimate terms, even when we are not aware of the actual instances of communication.

This process also has its mirror: We are listening to strangers all the time as well. Nothing that takes place anywhere in the world and is communicated on the Internet is at a remove any longer. Just as everyone on the Internet is a potential recipient and transmitter of our signals, we, too, are stations for the reception and relay of other people's messages. This constancy of connection to the nervous systems of billions of others comes with its own consequences.

No one can be immune to the storms that shake the world today. What happens down our streets becomes as present in our lives as what happens down our modems. This makes us present in vital and existential ways to what might be happening at a great distance, but it also brings with it the possibility of a disconnect with what is happening around us or near us if it happens not to be online.

This is especially true of things and people who drop out, or are forced to drop out, of the network, or who are in any way compelled not to be present online. This foreshortening (and occasional magnification) of distances and compression of time compels us to think in a more nuanced way about attention. Attention is no longer a simple function of things that are available for the regard of our senses. With everything that comes to our attention, we have to now ask, "What obstacles did it have to cross to traverse the threshold of our considerations?" and while asking this, we have to understand that obstacles to attention are no longer a function of distance.

The Internet also alters our perception of duration. Sometimes, when working on an obstinately analog process, such as the actual fabrication of an object, the internalized shadow of fleeting Internet time in our consciousness makes us perceive how the inevitable delays inherent in the fashioning of things (in all their messy "thingness") cause us to appreciate the rhythms of the real world. In this way, the Internet's pervasive co-presence with real-world processes ends up reminding us of the fact that our experience of duration is now a layered thing. We now have more than one clock, running in more than one direction, at more than one speed.

The simultaneous availability of different registers of time, made manifest by the Internet, also creates a continuous archive of our online presences and inscriptions. A message is archived as soon as it is sent. The everyday generation of an internal archive of our work and the public archive of our utterances (on online discussion lists and on Facebook) means that nothing is a throwaway observation anymore, not even a throwaway observation. We are all accountable to, and for, the things we have written in e-mails or posted on online fora. We have yet to get a full sense of what this actually implies in the long term. The automatic generation of a chronicle and a history colors the destiny of all statements. Nothing can be consigned to amnesia, even though it may appear insignificant. Conversely, no matter how important a statement may have seemed when it was first uttered, its significance is compromised by the fact that it is ultimately filed away as just another datum, a pebble in a growing mountain range.

Whoever maintains an archive of his practice online is aware of the fact that he alters the terms of his visibility. Earlier, one assumed invisibility to be the default mode of life and practice. Today visibility is the default mode, and one has to make a special effort to withhold any aspect of one's practice from visibility. This changes the way we think about the relationship between the private memory of a practice and its public presence. It is not a matter of whether this leads to a loss of privacy or an erosion of spaces for intimacy; it is just that issues such as privacy, intimacy, publicity, inclusion, and seclusion are now inflected very differently.

Finally, the Internet changes the way we think about information. The fact that we do not know something that exists in the extant expansive commons of human knowledge can no longer intimidate us into reticence. If we do not know something, someone else does, and there are enough ways around the commons of the Internet that enable us to get to sources of the known. The unknown is no longer that which is unavailable, because whatever is present is available on the network and so can be known, at least nominally if not substantively. A bearer of knowledge is no longer armed with secret weapons. We have always been autodidacts, and knowing that we can touch what we do not yet know and make it our own makes working with knowledge immensely playful and pleasurable. Sometimes a surprise is only a click away.

Dowsing Through Data

Xeni Jardin

Tech culture journalist; partner, contributor, coeditor, Boing Boing; *executive producer, host,* Boing Boing Video

I travel regularly to places with bad connectivity: small villages, marginalized communities, indigenous lands in remote spots around the globe. Even when it costs me dearly, as with a spendy satphone or gold-plated roaming charges, my search itch, my tweet twitch, my e-mail toggle—those acquired instincts—now persist.

The impulse to grab my iPhone or pivot to the laptop is now automatic when I'm in a corner my wetware can't get me out of. The instinct to reach online is so familiar now, I can't remember the daily routine of creative churn without it.

The constant connectivity I enjoy back home means never reaching a dead end. There are no unknowable answers, no stupid questions. The most intimate or not-quite-formed thought is always seconds away from acknowledgment by the great "out there."

The shared mind that is the Internet is a comfort to me. I feel it most strongly when I'm in those faraway places, tweeting about tortillas or volcanoes or vodoun kings, but only because in those places so little else is familiar. But the comfort of connectivity is an important part of my life when I'm back on more familiar ground and take it for granted.

The smartphone in my pocket yields more nimble answers than an entire paper library, grand and worthy as the library may be. The paper library doesn't move with me throughout the world. The knowledge you carry with you is worth more than the same knowledge when it takes more minutes, more miles, more action steps to access. A tweet query, a Wikipedia entry, a Googled text string, all are extensions of the internal folding and unfolding I used to call my own thought. But the thought process that was once mine is now ours, even while in progress, even before it yields a finished work.

That's how the Internet changed the way I think. I used to think of thought as the wobbly, undulating trail I follow to arrive at a final, solid, completed work. The steps you take to the stone marker at the end. But when the end itself is digital, what's to stop the work from continuing to undulate, pulsate, and update, just like the thought that brought you there?

I often think now in short bursts of thought, parsed out 140 characters at a time, or blogged in rough short form. I think aloud and online more, because the call and response is a comfort to me. I'm spoiled now, spoiled in the luxury of knowing there's always a ready response out there, always an inevitable ping back. Even when the ping back is sour or critical, it comforts me. It says, "You are not alone."

I don't believe there's such a thing as too much information. I don't believe Google makes us dumber, or that prolonged Internet fasts or a return to faxes is a necessary part of mind health. But data without the ability to divine is useless. I don't trust algorithm the way I trust intuition: the art of dowsing through data. Once, wisdom was measured by memory, by the capacity for storage and processing and retrieving on demand. But we have tools for that now. We made machines that became shared extensions of mind. How will we define wisdom now? I don't know, but I can ask.

Bleat for Yourself

Larry Sanger

Cofounder of Wikipedia and Citizendium

The instant availability of an ocean of information has been an epoch-making boon to humanity. But has the resulting information overload also deeply changed how we think? Has it changed the nature of the self? Has it even—as some have suggested—radically altered the relationship of the individual and society? These are important philosophical questions, but vague and slippery, and I hope to clarify them.

The Internet is changing how we think, it is suggested. But *how* is it, precisely? One central feature of the "new mind" is that it is spread too thin. But what does *that* mean?

In functional terms, being spread too thin means we have too many Websites to visit, we get too many messages, and too much is "happening" online (and in other media) that we feel compelled to take on board. Many of us lack effective strategies for organizing our time in the face of this onslaught. This makes us constantly distracted, unfocused, and less able to perform heavy intellectual tasks. Among other things, or so some have confessed, we cannot focus long enough to read whole books. We feel unmoored, and we flow along helplessly wherever the fast-moving digital flood carries us.

We do? Well—*some* of us do, evidently.

Some observers speak of "where we are going" or of how "our minds" are being changed by information overload, apparently despite ourselves. Their discussions make erstwhile free agents mere subjects of powerful new forces, and the only question is where those forces are taking us. I don't share the assumption here. When I saw the title of Nick Carr's essay in the *Atlantic*, "Is Google Making

Us Stupid?" I immediately thought, "Speak for yourself." Discussions such as Carr's assume that intellectual control has already been ceded—but that strikes me as being a cause, not a symptom, of the problem Carr bemoans. After all, the exercise of freedom requires focus and attention, and the ur-event of the will is precisely focus itself. Carr unwittingly confessed what is for too many of us a moral failing, a vice; the old name for it is *intemperance* (in the older, broader sense, contrasted with *sophrosyne*, moderation or self-control). And, as with so much of vice, we want to blame it on anything but ourselves.

Is it really true that we no longer have any choice but to be intemperate in how we spend our time, in the face of the temptations and shrill demands of networked digital media? New media are not *that* powerful. We still retain free will, which is the ability to focus, deliberate, and act on the results of our deliberations. If we want to spend hours reading books, we still possess that freedom. Only philosophical argument could establish that information overload has deprived us of our agency; the claim, at root, is philosophical, not empirical.

My interlocutors might cleverly reply that in the age of Facebook and Wikipedia we do still deliberate, but collectively. In other words, for example, we vote stuff up or down on Digg, del.icio.us, and Slashdot, and then we might feel ourselves obligated—if we're participating as true believers—to pay special attention to the top-voted items. Similarly, we attempt to reach "consensus" on Wikipedia, and—again, if participating as true believers—we endorse the end result as credible. To the extent that our time is thus directed by social networks engaged in collective deliberation, we are subjugated to a "collective will," something like Rousseau's notion of a general will. To the extent that we plug in, we become merely another part of the network. That, anyway, is how I would reconstruct the collectivist-determinist position opposed to my own individualist-libertarian one.

But we obviously have the freedom not to participate in such networks. And we have the freedom to consume the output of such networks selectively and while holding our noses—to participate, we needn't be true believers. So it is very hard for me to take the "Woe is us, we're growing stupid and collectivized like sheep" narrative seriously. If you feel yourself growing ovine, bleat for yourself.

I get the sense that many writers on these issues aren't much bothered by the unfocusing, de-liberating effects of joining the hive mind. Don Tapscott has suggested that the instant availability of information means we don't have to memorize anything anymore—just consult Google and Wikipedia, the brains of the hive mind. Clay Shirky seems to believe that in the future we will be enculturated not by reading dusty old books but in something like online fora, plugged into the ephemera of a group mind, as it were. But surely, if we were to act as either of these college teachers recommend, we'd become a bunch of ignoramuses. Indeed, perhaps that's what social networks are turning too many kids into, as Mark Bauerlein argues cogently in *The Dumbest Generation.* (For the record, I've started homeschooling my own little boy.)

The issues here are much older than the Internet. They echo the debate between progressivism and traditionalism found in the philosophy of education: Should children be educated primarily so as to fit in well in society, or should the focus be on training minds for critical thinking and filling them with knowledge? For many decades before the advent of the Internet, educational progressivists have insisted that in our rapidly changing world knowing mere facts is not what is important, because knowledge quickly becomes outdated; rather, being able to collaborate and solve problems together is what is important. Social networks have reinforced this ideology by seeming to make knowledge and judgment collective functions. But the progressivist position on the importance of learning facts and training individual judgment withers under scrutiny, and, *pace* Tapscott and Shirky, events of the last decade have not made it more durable.

In sum, there are two basic issues here. Do we have any choice about ceding control of the self to an increasingly compelling hive mind? Yes. And should we cede such control, or instead strive, temperately, to develop our own minds very well and direct our own attention carefully? The answer, I think, is obvious.

BOOKS BY JOHN BROCKMAN

CULTURE

978-0-06-202313-1 (paperback)

Coming Summer 2011

THE MIND

978-0-06-202584-5 (paperback)

Coming Summer 2011

IS THE INTERNET CHANGING THE WAY YOU THINK?

ISBN 978-0-06-202044-4 (paperback)

THIS WILL CHANGE EVERYTHING

ISBN 978-0-06-189967-6 (paperback)

WHAT HAVE YOU CHANGED YOUR MIND ABOUT?

ISBN 978-0-06-168654-2 (paperback)

WHAT ARE YOU OPTIMISTIC ABOUT?

ISBN 978-0-06-143693-2 (paperback)

WHAT IS YOUR DANGEROUS IDEA?

ISBN 978-0-06-121495-0 (paperback)

WHAT WE BELIEVE BUT CANNOT PROVE

ISBN 978-0-06-084181-2 (paperback)

Visit www.AuthorTracker.com
for exclusive information on your favorite HarperCollins authors.

Available wherever books are sold, or call 1-800-331-3761 to order.